STO

ACPL ITEM
DISCARDED

670
CHORAFAS, DIMITRIS N.
EXPERT SYSTEMS IN
MANUFACTURING

IN MANUFACTURING

**DO NOT REMOVE
CARDS FROM POCKET**

ALLEN COUNTY PUBLIC LIBRARY
FORT WAYNE, INDIANA 46802

You may return this book to any agency, branch,
or bookmobile of the Allen County Public Library.

DEMCO

EXPERT SYSTEMS IN MANUFACTURING

Dimitris N. Chorafas

With Forewords by Steven R. Belmont
and Toshiro Terano

AUTOMATION IN MANUFACTURING SERIES

VNR VAN NOSTRAND REINHOLD
_____ New York

Allen County Public Library
Ft. Wayne, Indiana

Copyright © 1992 by Van Nostrand Reinhold

Library of Congress Catalog Card Number 91-26023
ISBN 0-442-00827-9

All rights reserved. No part of this work covered by the
copyright hereon may be reproduced or used in any form or
by any means—graphic, electronic, or mechanical, including
photocopying, recording, taping, or information storage and
retrieval systems—without written permission of the
publisher.

Manufactured in the United States of America

Published by Van Nostrand Reinhold
115 Fifth Avenue
New York, New York 10003

Chapman and Hall
2-6 Boundary Row
London, SE1 8HN, England

Thomas Nelson Australia
102 Dodds Street
South Melbourne 3205
Victoria, Australia

Nelson Canada
1120 Birchmount Road
Scarborough, Ontario M1K 5G4, Canada

16 15 14 13 12 11 10 9 8 7 6 5 4 3 2 1

Library of Congress Cataloging-in-Publication Data
Chorafas, Dimitris N.
 Expert systems in manufacturing / Dimitris N. Chorafas.
  p. cm.–(Automation in manufacturing)
 Includes index.
 ISBN 0-442-00827-9
 1. Production engineering–Data processing. 2. Expert systems
(Computer science) 3. Computer integrated manufacturing systems.
I. Title. II. Series.
TS176.C49 1991
670'.285–dc20   91-26023
                                                          CIP

*To Professor Toshiro Terano
and Steven R. Belmont
For their leadership in systems engineering*

Automation In Manufacturing Series

Sourcebook of Automatic Identification, Russ Adams
Perspectives on Radio Frequency Identification, Ron Ames
Handbook of Bar Coding Systems, Harry E. Burke
Automating Management Information Systems: Vol. 1. Principles of Barcode Applications, Harry E. Burke
Automating Management Information Systems: Vol. 2. Barcode Engineering and Implementation, Harry E. Burke
Working Toward JIT, Anthony Dear
Electronic Data Interchange: A Total Management Guide, Margaret Emmelhainz
Automatic Identification: Making It Pay, Kevin Sharp
Competitive Manufacturing through Information Technology: An Executive Guide, John Stark
Inventory Accuracy, Jan Young
Knowledge Engineering, Dimitris N. Chorafas
Expert Systems in Manufacturing, Dimitris N. Chorafas

Contents

Foreword by Steven R. Belmont / **xiii**
Foreword by Toshiro Terano / **xv**
Preface / **xvii**
Part 1: The Broadening Perspective of Knowledge Engineering in the Manufacturing Industry / **1**

1. Planning for Excellence in Manufacturing / **3**

1. Introduction / **3**
2. The Manufacturing Industry's New Era / **4**
3. Assisting Man's Physical and Mental Work / **7**
4. The Need for Structural Reform / **10**
5. Organizational Prerequisites and Technological Solutions / **13**
6. The Enviable Record of the Forerunners / **15**
7. A Factory of the Past or of the Future? / **17**
8. The Integration of Technological Investments / **19**

2. Revitalizing Factory Automation / **21**

1. Introduction / **21**
2. A Long Hard Look at Factory Automation / **22**
3. An Integrative Approach to Manufacturing Excellence / **25**
4. The Role of Intelligent Communications Links / **28**

5. Establishing a System Architecture / **31**
6. A Case Study at General Electric / **34**
7. Requirements for a Transition Policy / **35**
8. Learning by Example and Just-in-Time Inventories / **37**

3. Knowledge Engineering in the Manufacturing Industry / **40**

1. Introduction / **40**
2. The Competitor's Impact on Our Operations / **41**
3. The Expanding Domain of Expert Systems in Manufacturing / **44**
4. A Golden Horde of Implementation Examples / **46**
5. ACE and DELTA / **48**
6. Contrasts and Similarities in Expert Systems Applications / **50**
7. Placing Emphasis on Diagnosis and Fault Isolation / **52**

4. Practical Examples with Expert Systems / **55**

1. Introduction / **55**
2. Expert Systems Projects by Bechtel Corporation / **57**
3. Use Your Intellect to Get Results / **59**
4. Expert Systems Projects by Inference Corporation / **62**
5. Multimedia Expert Systems for Topological Applications / **64**
6. Implementation Areas That Have Given Commendable Results / **67**

5. AI in Network Design and in Power Engineering / **70**

1. Introduction / **70**
2. Expert Systems for Network Design at Bolt Beranek Newman (BBN) / **72**
3. Global Network Design Through an Expert System / **74**
4. Payoffs from AI Solutions in the Telecommunications Domain / **77**
5. Expert Systems at the Lawrence Livermore National Laboratory (LLNL) / **80**
6. Alarm Filtering, Emergency Control, and Other Power Production Expert Systems / **84**

6. Knowledge Engineering at Digital Equipment Corporation and at IBM / **87**

1. Introduction / **87**
2. DEC's Thrust into Knowledge Engineering / **88**
3. The Concept of an Expert Configurer / **92**
4. XCON's Documented Benefits / **94**
5. A Development Timetable and its Challenges / **96**
6. IBM's Thrust in AI / **99**
7. The Capital Asset Expert System (CASES) / **101**

7. The Impact of Second Generation Solutions / 105

1. Introduction / **105**
2. The Challenge of Fuzzy Engineering / **106**
3. The Contribution of Cognitive Sciences / **112**
4. Pattern Recognition and Oil Exploration / **114**
5. Pattern Analysis and Metaknowledge / **116**
6. The Search for Enriched Approaches to Database Usage / **118**

8. Advances in Robotics / 122

1. Introduction / **122**
2. Trends in the Robotization of Industry / **123**
3. Artificial Intelligence in Robot Control / **125**
4. Towards Autonomous Vehicles and Microminiaturization / **127**
5. The Enlarging Domain of Field Robots / **129**
6. Practical Applications of Mobile Robots / **131**
7. Artificial Vision and Fuzzy Engineering / **133**
8. High-Level Vision Systems and Sixth Generation Computers / **135**

Part 2: Computer-Aided Design, Production Scheduling, Quality Management, Marketing, and Computer-Integrated Manufacturing / 139

9. A New Generation of CAD Solutions / 141

1. Introduction / **141**
2. Towards Second Generation CAD / **142**
3. Approaches to the Use of an Intelligent CAD System / **146**
4. Solutions to be Obtained through High Technology / **150**
5. Managing our Analytic and Synthetic Skills / **152**
6. Component Parts of a Valid Approach to Expert Systems / **154**
7. Profile of Expertise in Engineering Design / **157**

10. Providing the Engineering Designer with Insight and Foresight / 160

1. Introduction / **160**
2. Utilizing an Acquired Experience Effectively / **161**
3. Enriching the Simulator Through AI / **164**
4. Benefits from the Use of Expert Systems in Product Design / **167**
5. A Possible Study of Diesel Engines / **169**
6. Approaches to Image Analysis / **173**

11. Expert Systems in Production Planning / **176**

1. Introduction / **176**
2. The Knowledge-Based Manufacturing Enterprise / **177**
3. Production Planning Through Expert Systems / **181**
4. AI Implementation Examples at the Production Floor / **184**
5. Vagueness and Uncertainty in Production Planning / **186**
6. Salient Problems in Production Control / **189**
7. Using Knowledge Engineering for Process and Labor Management / **191**

12. An Infrastructure for Production Scheduling / **194**

1. Introduction / **194**
2. Constraints in Stabilizing a Production Schedule / **196**
3. Vital Supports in Production Management / **199**
4. Mathematical Programming and Scheduling Methods / **202**
5. The Assignment of Responsibilities / **204**
6. Improving the Overall Efficiency Through Input/Output Solutions / **208**
7. Distributed Negotiations for Salient Problems in Production Scheduling / **210**

13. The Challenge of Quality Assurance / **213**

1. Introduction / **213**
2. Gaining Leadership in Quality Assurance / **214**
3. Artificial Intelligence in Quality Control / **217**
4. Expert Systems as Fault Finders / **220**
5. Expert Assistance in Designing and the Quality Assurance Challenge / **223**
6. The Process of Online Diagnosis / **225**
7. Using Evidential Reasoning in Failure Detection / **227**
8. Assuring the Success of Knowledge Engineering in Diagnosis / **229**

14. Selling the Products That We have Made / **232**

1. Introduction / **232**
2. Marketing Strategy and Corporate Objectives / **233**
3. AI in the Service of Sound Policies / **237**
4. Technology and the Story of Marketing / **239**
5. Sales, Distribution and Field Support / **242**
6. Merging Telecommunications and Expert Systems / **244**
7. Practical AI Applications to Assist the Marketing Effort / **246**

15. Computer-Integrated Manufacturing / 252

1. Introduction / **252**
2. The Challenge of Disconnected Areas in Manufacturing Operations / **253**
3. A Critical Look at Integrative Requirements / **257**
4. Keeping Track of Project Status / **259**
5. Rethinking Our Cost Control Policies / **260**
6. The Cost of Staying in Business / **264**
7. CIM as the Key to Manufacturing Competitiveness / **267**

16. Projecting a Rational CIM Architecture / 270

1. Introduction / **270**
2. The Architecture of the Flexible Manufacturing Enterprise / **271**
3. Concurrent Engineering, an Example of CIM Architecture / **274**
4. Tuning the CIM Architecture to the Product Line / **277**
5. Obtaining Better Coordination—From Product Design to Production Scheduling / **281**
6. Observing Security Requirements / **283**
7. CIM Architecture and Market Competitiveness / **285**

Index / **289**

Foreword

Are ES (Expert Systems) and AI (Artificial Intelligence) ready to become industrial-household computing buzzwords, like CIM, SPC, TQM, JIT, MRP. For manufacturing to become truly competitive, let's hope ES and AI are successful implementations, not just buzzwords. While it is true that almost anyone in the manufacturing sector is familiar with CIM, SPC and other acronyms and what they represent, and most professionals have applied one technique or another (occasionally unknowingly), implementation has been flawed, cultural resistance encountered, and success limited.

One possible reason for any (perceived) lack of success is the fact that these other "wonder tools" typically ignored the non-manufacturing aspects of an enterprise. These key areas–customer service, strategic planning, and others–are vital to an organization's health. Commitment to the future through strategic planning ensures the long term vitality of any company, just as expanding the customer base has the greatest impact on its near-term health. These bear equal weight to the "Manufacturing" aspects of a company. A well-produced product with no market cannot be considered a success. In the long run, the success of any tool must be measured by its direct impact on the bottom line. The most sound mathematical or computer aid will not even be considered unless it can show a positive impact on profitability–both short and long term.

Most people have some familiarity with at least some concepts of ES and AI, especially "knowledge base" and "rules." But often familiarity does not go beyond these general concepts, and to date, successes have not been exploited for the benefit of any company as a whole.

If however, both *manufacturing* and *non-manufacturing* areas in an enterprise can be shown to benefit from application of expert systems and artificial intelligence, conversion and acceptance throughout the company will be more rapid than ever before.

Does an ES tool exist that can enhance the performance of non-mainstream qualitiative areas such as sales and customer service? Are there real examples of well known companies benefitting from using expert systems, now?

Expert Systems in Manufacturing gives a resounding yes to both these questions and more. Dr. Chorafas gives examples of all areas of the enterprise to benefit from ES and AI applications. One of the best examples is of a computer manufacturer realizing unimaginable returns in its customer service/field function using just such a tool. Another is of a computer vendor using ES to configure and design a customer's system. Many examples of implementations of "fuzzy logic" are included, in applications to functions not normally associated with computerization.

In presenting new information Dr. Chorafas does not exort us to relie on intuitive reasons for justifying new technical implementations. Instead, he illustrates and justifies by example applications of AI and ES throughout an enterprise as a boost to profits and efficiency. Bettering the bottom line is the best testimony to adoption of new tools. This book will show you how it has been done, and how you can do it too.

Steven R. Belmont
Manager, Systems Integration Services
Burr-Brown Corporation

Foreword

It is an honor for me to write a foreword for Professor Chorafas's new book. He is one of my most respected friends. I have studied and taught at the Transportation Research Institute, Tokyo Institute of Technology, Hosei University, and am now an executive director of the Laboratory for International Fuzzy Engineering Research. During the past 40 years my field of expertise has shifted from automation to systems engineering, and most recently, to fuzzy engineering. This shift of interest may appear to be inconsistent, but it is actually quite natural.

The need for automation is very clear. Every business and industry wishes to reduce the human load and to increase efficiency without decreasing quality. This effect of automation is greater when the business is larger and more complex. The goal of automation for these complex systems is to keep conditions constant and also includes the automatic start-and-stop, optimum control, safety control, noninteractive control, etc.

To reach these goals, the initial design and management of the systems is sometimes more important than the automation of the existing facilities. Systems engineering is a methology which has been developed to plan, design, develop, construct, manage, and operate large-scale systems rationally. The able use of the computer plays an essential role in systems engineering and this approach is now indispensable in every line of industry.

When considering the development of systems engineering some difficult problems related to human factors arise. For example, problems such as human error, the strategy of the decision-maker, evaluation by the designer, and publicity. Many of these issues have not been previously considered in the engineering field and there is no effective methodology for doing so, except for fuzzy engineering.

My interest has been increased whenever I found a new problem. When we plan for automation we must consider the integrated effect of the whole factory, not just the automation of a machine. Then careful consideration must be given as the emphasis shifts from a local problem to an integrated one, that is, from operational considerations to those of management, and from controller to computer.

Another change is the recognition of the human role in the system. In the old control engineering it was considered ideal for machines to perform every job, replacing the human worker. It is now realized that workers must support machines to supplement the incompleteness of automation. This is curious, because the machine appears to be the master which humans serve. The man-machine system, on the contrary, is a new concept in which humans make the final decision and machines support it. Humans can fully display this ability in managing such systems. Artificial intelligence, or expert systems, are very powerful tools which assist human thought. This suggests that automation changes its role from the support of physical jobs to the support of mental effort performed by humans.

In accordance with the development of technology, the manufacturing system itself should be changed. However people are very conservative, and they resent any changes in organization or strategy, even if they understand the need. One of the most important things to consider when undertaking changes is how to adapt peoples minds to the rapid progress of computer technology. It is helpful for people to recognize not only the present situation, but also the long term trend of technology.

This book, I think, is very farsighted and also practical. The author explains the most modern technology from both narrow and wide viewpoints, but he still keeps in sight the user and the use of technology. Therefore the reader will easily understand the real need for expert systems and how to use them in real situations. I hope the reader will learn from this book not only the techniques of knowledge engineering but also the concept of the man-machine system. This book will prove to be very valuable for all engineers and managers who aspire to improve manufacturing systems.

Dr. Toshiro Terano
Professor, Hosei University
Executive Director, Laboratory for
International Fuzzy Engineering Research

Preface

A *manufacturing enterprise* is generally understood as a system of design and production units. These are groups of *men* and *machines*—the former equipped with appropriate tools, the latter with the right software. *Production* means any kind of work necessary in the broadest sense: developing, engineering, planning, constructing, testing, buying, selling, controlling and investing.

The management of increasingly complex manufacturing processes underlines the need for assembling and exploring a massive knowledge base spanning many areas of human know-how. In the long run, this means a kind of crossover primarily from manual skills to a very sophisticated knowledge of products and processes.

The exploitation of such knowledge-based systems presumes the existence of a manufacturing model broad enough to allow a clean distinction between production issues and how *our company* relates to the market to which it appeals. This is the role *expert systems* are called to play. Hundreds of them have been developed and implemented in manufacturing environments, and they have been used for years with a considerable measure of success.

This book is written for manufacturing engineers, industrial engineers and design engineers who would like to obtain good background of the way expert systems have been *applied* in the operations of the manufacturing industry. It also addresses itself to managers who care to focus on the *benefits* artificial intelligence (AI) presents as well as the *results* to be obtained through computer-integrated manufacturing.

The book divides into two parts each composed of eight chapters. Part One addresses itself to the broadening prospects of knowledge engineering in the fabrication industries, starting with the need to *plan for excellence* in manufacturing (Chapter 1).

The focal point of Chapter 2 is the ways and means for revitalizing factory automation. Chapter 3 outlines the role knowledge engineering can play in this process. Chapter 4 subsequently offers a great number of practical examples of expert systems' use in the manufacturing industry.

The goal of embedding AI in a production process is not to operate a perfect system, but rather, to develop solutions that are agile, adaptable and robust. This strengthens our defense capabilities *against competition* and also creates the preconditions for successfully facing the growing requirements of the market.

Along this line of reasoning, Chapter 5 explains how AI has been put to use in domains such as network design and power engineering. Examples are taken from Bolt, Beranek, Newman as well as from the Lawrence Livermore National Laboratory. Chapter 6 focuses on the usage of knowledge engineering in IBM and DEC, the two top computer companies.

There is also a new generation of expert systems for which we have to account. They consist of fuzzy engineering constructs, neural networks, as well as the able experimentation on intelligent database approaches, as discussed in Chapter 7. Companies are now actually working on these subjects and therefore new solutions are coming.

In terms of implementation, the manufacturing industry has also benefited from a golden horde of robots. Chapter 8 talks about trends in the robotization of industry and the growing domain of artificial vision.

A fully intelligent enterprise is an unattainable idealization since it seems to be impossible to give an exhaustive set of attributes characterizing every single operation in its finest detail. But organizations use artifacts that can benefit from modeling and expert-system support, thus helping to spread logic and experience throughout the organization.

In terms of a methodological and consistent approach, able solutions to knowledge representation and inference techniques are realizable and practically useful. By and large, AI models are not aimed to replace but to extend and to complete the existing experience in a manufacturing enterprise.

Part Two capitalizes on these concepts, starting with a new generation of computer-aided design (CAD) solutions presented in Chapter 9. This is followed, in Chapter 10, by a description of the means necessary to provide the engineering designer with insight and foresight in his work.

The use of expert systems in production planning is the subject of Chapter 11. As other sectors of the manufacturing company such as sales forecasting become more advanced, we find that the production planning solutions we used more or less adequately yesterday become too limited if not outright unacceptable. New, more powerful approaches are needed for scheduling—including a streamlined infrastructure—as Chapter 12 suggests.

The focal point of Chapter 13 is quality assurance. More than ever, the battlefield of the 1990s will be in the high quality domain where the competitive edge lies. The use of practical expert systems for marketing, sales, distribution and field support is explained in Chapter 14. Integrating marketing, product innovation and production processes is one of the most promising strategic applications within a modern organization.

Chapters 15 and 16 concentrate on computer-integrated manufacturing. The first talks about the cultural and technical challenges posed by CIM, the former being much greater than the latter. The second suggests an effective architectural solution for the able implementation of computer-aided manufacturing approaches.

For some types of operational activities within the realm of the manufacturing industry, there are strategies for resource allocation and problem handling that can be applied in almost all companies. Computer-aided manufacturing is one of them. It is a relatively new technology with much to offer if we know how to implement it correctly. As a number of studies suggest, this is the way to go regarding *investments* in the 1990s. The Industrial Bank of Japan, for instance, estimates that between 60 and 70 percent of capital budgets in the 1990s will be spent on imaginative *labor-saving* projects as well as on developing and supporting flexible *new products*. This should be our guideline for all future investments in organization and in technology.

The writer feels indebted to a great number of manufacturing executives, bankers and technologists who contributed ideas and reviewed and commented on many parts of this book. To Eva-Maria Binder goes the credit for the artwork, typing and index.

Dr. Dimitris N. CHORAFAS
Valmer and Vitznau

EXPERT SYSTEMS IN MANUFACTURING

part 1

THE BROADENING PERSPECTIVE OF KNOWLEDGE ENGINEERING IN THE MANUFACTURING INDUSTRY

1

Planning For Excellence In Manufacturing

1. INTRODUCTION

In the old mechanical fabrication plant, labor costs accounted for 75 percent or more of the product cost; the 15 percent was for materials. By contrast, in today's electronic products plants, between 55 percent and 75 percent of cost embedded in a product is for parts and materials; labor accounts for a small fraction of the production cost and is on the way down.

In fact, the cost of direct labor in a typical high technology company seldom exceeds 10 percent of the product cost and it is expected to fall under 5 percent, about the same as depreciation. In some industries this has already taken place. The typical labor content of a personal computer today is about 2 percent. Many companies could drive it lower but simply do not bother since it hardly matters in terms of the overall cost.

This reference, however, is not valid in what regards *indirect* factory costs. These arise all the way from process engineering, and software development to materials control, quality assurance and maintenance—and they are growing rapidly. In many firms they now equal *from five to ten times* their direct labor costs.

What is now incorrectly called "indirect costs" is necessary for many reasons beginning with product and process innovation and ending with quality assurance and industrial engineering to squeeze out direct costs. This is how although in America labor costs per hour rose 16 percent between 1987 and 1989—to $29.50 per hour from $25.40—the direct labor cost of a GM vehicle rose a scant 1.7 percent.

This change is largely responsible for dramatic reductions in the manufacturing work force. Not only has the share of labor and materials in the cost of goods sold been reversed, but companies are also increasingly switching from large centralized metal working and assembly plants to smaller, distributed production and processing units. At the production floor itself, information technology makes better coordination feasible:

- Robots automate the formerly labor-intense operations,
- Automatic scheduling systems control shorter lines for making goods,
- Buffer stocks between manufacturing stations (just-in-time inventories) should to be sharply reduced,
- Quality control processes should aim to be better tuned to product assurance than they were in the past, and
- Steady emphasis on cost control to become and remain a low-cost producer.

In other words, robotics not only displaces labor and alters the classical manufacturing chores, but it also makes feasible significant cost improvements through computer-based production planning and control—including scheduling, inventory management, quality control, and purchasing activities.

A *new culture* has been born in manufacturing and this culture is the subject of the present book. The challenge is how to adopt this new culture; we will approach this subject by looking into what successful companies have done.

2. THE MANUFACTURING INDUSTRY'S NEW ERA

In July 1990, the US Department of Commerce identified twelve emerging technologies that will create an estimated $1 trillion by the year 2000. One of them is flexible computer-integrated manufacturing; a second advanced materials; a third artificial intelligence.

These three emerging domains work in synergy and together imply a number of prerequisites that have to be met. One of them is the ability to understand and prepare for the revolutionary capabilities of the new technologies. Short of that, these technologies will become poorly used and a lot more expensive than they need to be.

The 1990s do not only permit but also oblige us to develop new work methods, so that we develop a perspective of manufacturing activities different from the one we had in the past. To do so, we must develop a strategic plan that is both marketing and technology oriented, fine tuning it all the way as Figure 1-1 suggests.

A manufacturing company engages in strategic planning when it:

- Selects and defines its objectives,
- Determines the means required for achieving them, and
- Prepares for reaching those objectives systematically within stated time periods.

Fives years is usually regarded as the minimum period required for meaningful planning. A ten-to-fifteen-year period is more desirable. Plans have been projected as far as twenty years ahead of time. This too calls for a cultural change:

- If we do not change our culture and our ways of doing things, we will waste the small fortune we paid for the machines and for acquiring new skills and

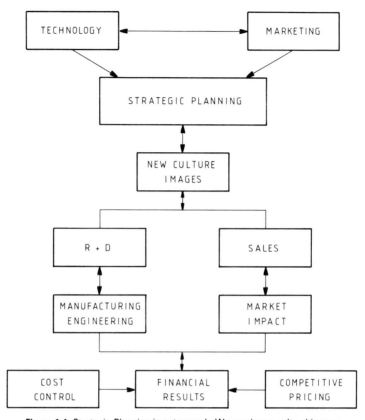

Figure 1-1. Strategic Planning is not enough. We need new cultural images.

logistical support; we will not get market results.
- If we do not control our costs, we will drive ourselves out of the market.

This statement is valid in regard to all our expenditures. We should not buy communications, computers and robotics—and we should not hire the rocket scientists with them—unless we are committed to getting the most out of these resources.

We have to organize to reach such goals and renewal has to be factory-wide. The sophistication of the procedures that govern automated machine operation magnifies the errors of faulty upstream processes and their effects on downstream decisions.

Manufacturing management has to recognize that results will not be obtained by throwing money at the problem. To emulate an expert machinist's talent for recognizing malfunctions, an automated system needs:

- Elaborate online databases,
- Expert systems incorporating the implicit rules of the skilled operator, and
- A scientific understanding of the technology itself.

Manufacturing engineering must provide the system sensors to detect deviations. Artificial intelligence (AI) enriched controllers must work realtime to interpret online signals and initiate corrective action. The computer must be able to analyze the microstructure of manufacturing at microsecond speed. Short of that, "the factory of the future" will be a disaster—both financially and technologically.

The ultimate *measure of success* involves the *quality* and *cost* of the final product—and this is the real reason why an enterprise exists:

- If the product that comes out of that plant is a dog, it will not matter how advanced the manufacturing system is.
- If the product is highly reliable and cost/effective, then the technology used will become a trendsetter for the industry.

The importance of this statement to industrial America is that after years of emphasizing marketing and finance, we are now at the point of giving manufacturing a central place in corporate strategic planning. A whole new generation of concepts comes out of this switch, but are *we* ready for them?

An effective answer calls for much more than words. In the manufacturing era of the 1990s, the manager's job is mainly taken up with interpreting *patterns:*

- Learning to think more like a systems professional,
- Taking the necessary steps to maximize the capabilities of the new technologies, and
- Controlling and integrating processes and products, by understanding the nuances of the new landscape.

A manufacturing industry manager who does not understand one or more parts of the automated factory processes, as well as the products being produced, finds it impossible to make the necessary tradeoffs—between cost, quality and schedule, for instance. Postmortem understanding is not enough; product and process procedures must be developed, taking into consideration all possible consequences from design to field testing.

The new manufacturing discipline calls for organizations geared for a tight integration of:

- Product design,
- Process engineering,
- Quality management, and
- Cost control.

Executives who preside over advanced manufacturing establishments must be cross-disciplinary, having a deep understanding of automated machinery, innovative designs, cost/effective manufacturing, software engineering, and telecommunications.

There is a different way of looking at the subject of leadership in manufacturing and in engineering. Although in the past we used the computer for areas that were procedurally well-defined and stable, we now use computers and communications to help us look into the future. This job is more difficult; therefore, a deeper study and considerable research should be undertaken to sharpen our tools and to help us focus on our goals.

3. ASSISTING MAN'S PHYSICAL AND MENTAL WORK

During the last 15 years, computers have developed at least ten-thousandfold in speed and power. In 1975, IBM's 370/145 reached the benchmark million instructions per second (MIPS), which subsequently became a metric of computer speed. In 1990, Thinking Machines' Connection Machine-2 had a peak power of 10,000 MIPS at a roughly equivalent cost to the 1 MIPS. Even greater enhancements are coming.

The challenge is not only to make the power available but rather to put it to proper use—deriving benefits from its existence. Short of productive results, more MIPS will not mean power but rather the Meaningless Indication of Processing Speed.

The purposeful and profitable use of processing power is becoming increasingly important as in a number of tasks which underpin manufacturing, from computer-aided design (CAD) to robotics, man-made systems have largely supplanted human beings in the aiming and direction of factory operations:

- Man-made systems have been making considerable strides in the processing of huge amounts of industrial information.
- It is literally true that more large scale computers have been built in any recent two-year period than in all the history of mankind up to that time.

Therefore, in the 1990s, it is expected that computers will become an even more important factor in the manufacturing landscape, take over much of the information processing activity currently handled by human beings, and eventually revolutionize the role workers play in the First World.

This conclusion is based on the observation that a large part of the activity of all industrial workers—blue and white collar—is concerned with processing information. From classical data processing to the automation of logical processes through knowledge engineering, there are overlapping phases and evolutionary strides that are becoming more and more closely interrelated. These have to do with the handling of man's:

- Physical (and largely muscular) work, and of his
- Professional, managerial and clerical or mental activities.

Table 1-1 shows some of the milestones characterizing developments in both domains. First, it emphasizes the transition we have experienced in physical work—from the tool to the automatic factory in which physical and mental work merge. Second, it underlines the aids that have become available for information work.

In an automatic factory the computer portion of a control process is mature and tends to appear as a system capable of making intricate decisions at high speeds, through artificial intelligence constructs. But computer power has its price and its benefits. During the last 20 years:

- Ten percent computer-processing power translated into, and
- A 2 percent reduction in clerks.

TABLE 1-1. Milestone Developments in Man's Physical and Mental Work

Aids in Physical Work

- *The Tool:* man provides the power and the know-how
- *The Power Tool:* man contributes the knowhow.
- *The Automatically Controlled Power Tool:* man provides pilot control and coordinating machines activities.
- *The Automatic Factory, or* Complete Process Control, *man sets up groups of machines automatically coordinated and controlled.*
- *The Abacus,* desk calculator, and cash register serve as simple aids to computing.
- *The Business Machine,* processes a mass of similar documents through wired-in instructions.
- *The Computer,* performs complete sequences of programmed business or mathematical calculations.
- *The AI-Enriched Computer,* with rules and metarules, establishes elaborate programs using a flexible approach.

Through the impact of computers and of knowledge engineering on all types of arithmetic, logical and text-based operations—mathematics, the sciences, engineering, manufacturing, business, industry and finance—productivity-boosting forces make themselves felt in every sector of industry. This is true from manufacturing and marketing to transport, distribution and banking. As a result, while production increases, employment in the First World's manufacturing industries peaks and declines steadily.

Some companies have capitalized on this fact and they have been able to reap very significant benefits. Others spend tremendous amounts on technology but get meager results. Unfortunately, the latter are more numerous than the former. While spending on information technology has been climbing, two factors have delayed its effect on productivity.

1. For many years, companies have run paper systems alongside automated ones although as experience with computers and computer security grows, those paper systems have become irrelevant and often destructive.
2. It took time for management to discover how information technology would change the company's ability to add value and profit to its products and services.

It has been a long, slow transition often due to the resistance encountered in the introduction of new methods and of new technology. By far, the largest share of such resistance is due to psychology rather than technology—and *training* has often proven instrumental in breaking it down. But not all companies appreciate the wisdom of investing in *human capital.*

On the technological side, the drive to automation in the manufacturing industry did not start with digital computers but with analog machines. While today digital computers dominate, we should not forget that analog computing functions still show up in controls in a big way. We must appreciate the analog machine's ability to process information, but there are limits to this process.

While conversion to the digital solution is necessary to capitalize on new technologies, all by itself it is not enough. Some functions of information technology are more desired (and more profitable) than others are.

The expert systems now coming into the manufacturing picture perform these functions better, representing a further extension of the automation techniques. As the information processing requirements become more sophisticated, and complex analysis or optimizing operations are required at each stage of control, nothing short of full-fledged AI solutions can enhance *our* company's competitiveness.

To appreciate this statement, we must return to the fundamentals. Technology is concerned with the systematic production of a class of artifacts. All specific efforts are characterized by two common features.

1. *The need to be well-organized.* This means that there should exist an implicit and explicit body of knowledge that provides a conceptual framework, within which we can design or model new theories and do a practical implementation of science's products.
2. *A system of conventions* that masters any mature technology, providing standards and governing development processes. Standards reflect the criteria for judging the performance and the quality of artifacts.

References 1 and 2 can be found throughout the successive stages of mechanization and of automation we have experienced since the 19th century's Industrial Revolution. Figure 1-2 reflects this issue by mapping key steps in the evolution of manufacturing technology in the post-World War II years.

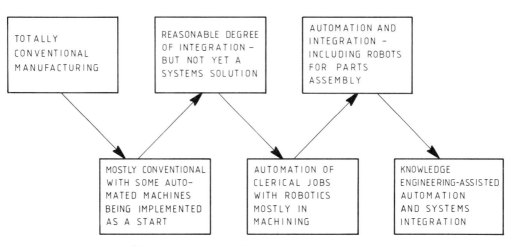

Figure 1-2. A steady evolution in manufacturing technology.

State of the art implementation requires the deployment of *knowledgeable* human resources as no integrative goal can be obtained without them. But since such resources are in great demand and in short supply, we are using knowledge engineering to assist in all phases of manufacturing activities—from product and process design, to the fabrication of components, their assembly, inventory control and quality inspection.

4. THE NEED FOR STRUCTURAL REFORM

The reasons for the slow pace at which fruits are being reaped from factory automation, in spite of high investments, do not necessarily lie in the inherent demands of the process. Rather, they can be found in the infrastructure that has become embedded in most manufacturing companies (and processes) during this century.

The reference just made includes the hierarchical structure of the organization as well as the survival of obsolete policies and inadequate procedures. Often such factors are so ingrained that they have become second nature. Yet:

- Traditional hierarchical attitudes and habits are incompatible with the requirements and capabilities of advanced technology,
- Piecemeal changes vastly underutilize the potential of new manufacturing investments, and
- Some managers, fearful of their position, are slow to appreciate the essence of the new technologies.

As section 2 has underlined, reform means nothing short of a new manufacturing culture and associative structures able to accommodate interactive, cooperative relationships—a kind of federated units, where the one impacts on the other through knowhow and not by means of flat orders. Reform also means changes in procedures all the way from production planning and inventory management to accounting, costing,* performance measurement and capital budgeting.

As technology for representing and manipulating production knowledge comes into being, old questions have to be asked in new ways. How is manufacturing information being handled in human thought processes? How does an industrial engineer explore a given diagnostic problem? How does the expert's problem-solving style differ from the novice's style? How well do different problem-solving styles work, and in what circumstances do different approaches give the best results?

Far from being academic, these questions define an information and decision support flow that is far from linear, as the reader can appreciate in Figure 1-3. Multiple feedbacks see to it that very sophisticated approaches are necessary to manage an environment of rapid innovation, growing complexity and stiffening competition.

Pressures mount at headquarters and at the plant to make substantial, not just marginal, improvements in performance. These include utilizing available capacity more efficiently, cutting costs, improving quality, and introducing new products at a faster pace. Because

---

* The process and metrics of cost control in industry.

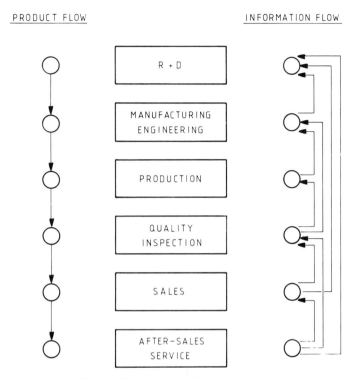

Figure 1-3. Distinguishing between product flow and information flow.

major improvements are the goal, no group can ask another to adjust its behavior without appreciating that such adjustments affect its own performance.

These references do apply to a vast number of firms—the more elementary ones, the specialized suppliers of components, and the largest firms. Today there are over 700,000 companies in the US alone engaged in manufacturing. The great majority of them (some 80 to 85 percent) have annual revenues under $5,000,000; therefore, restructuring should be no major deal—provided there is the wish to do it.

Not only the will for change but also a sound plan is necessary for restructuring the 100,000 larger companies directly concerned by the references being made, not just the Fortune 500 firms. All of them should feel the need to conserve capital and fuel growth, and to acquire the necessary functionality through the correct use of high technology.

The bottom line is that success in manufacturing today requires a new mind. In the wake of global competition, aggressive marketing, total quality control, just-in-time (JIT) inventories, and employee involvement have emerged as a basic frame of reference. Many changes are required for a manufacturing company to compete—not just building commitment among management but also establishing new job descriptions and their associated penalties and rewards.

Looking back at Figure 3 and the complex feedback mechanism that it suggests, we should properly appreciate that a great deal of change has come through four channels:

- Networked personal computers that make feasible multimedia communications,
- Large distributed databases that act as information repositories,
- Intelligent broadband networks interconnecting cooperating units in real time, and
- More widely implemented simulators and expert systems.

To exploit these facilities in an able manner, we must lay down a corporate architecture to assure that within *our* organization everybody's new automation project will work with everybody else's. This underlines the need for integration.

As the work of systems design and implementation becomes decentralized, we would require more and more of an integrative architecture* in order to tap the enthusiasm and expertise of those doing the jobs the system is meant to help. Working along this line of reasoning, Japan's NTT (whose activities roughly correspond to those of AT&T) developed the Multivendor Integration Architecture (MIA).

The concept of MIA is simple: it provides a software reference level that permits multivendor procurement policy through the assurance of compliance to the software standard. The work started in 1988 and the detailed specifications were published in February 1991 in partnership with IBM, DEC, Hitachi, Fujitsu and NEC.

After completion of the conceptual phase of the MIA project, the partners realized that flexibility is very important, requiring changes in the way individual systems are designed and built around a common software standard. This is true all the way from macroengineering to the development of programming products that are portable.

The traditional way of looking at systems design, known as the *waterfall model,* claimed that like a tangible machine, software evolves through several developmental stages of design and implementation until, one day, it is finished. Work done after that is called maintenance which often ended by consuming three quarters of the resources of systems specialists. The backlog for even simple new projects stretched out for years.

Part of the solution to this problem involves recognizing that in the manufacturing industry, as anywhere else, systems evolve. Hence new software solutions are needed:

- With expert systems we have *sustenance,* rather than maintenance. A flexible AI construct evolves throughout its lifecycle.**
- Implicit in this is the idea that computer software is never really finished; it merely passes through successive cycles of conceptual definition, design, implementation, testing and improvement.

The foremost specialists in this profession understand that *object-oriented* design methods help them get a grip on system complexity, and make change more manageable. Together with layered architectures, they enable solutions to be built in modules so that some functions can be altered and rewritten without changing anything else.

---

* See also D.N. Chorafas, Systems Architecture and Systems Design, McGraw-Hill, New York, 1989
** Sustenance differs from maintenance for two reasons: because of the *lifecycle* concept it introduces, and because it incorporates a process of steady improvement.

Forward-looking manufacturing companies are also building up inventories of the skills and knowledge their employees need, devising ways to enrich such inventories by industrializing and reproducing scarce knowledge. This is what *implementing* expert systems means.

Companies are developing rich databases that help people with problems get in touch with stored information and knowhow. When we talk of structural reform in manufacturing, we therefore make reference to a multilayered approach: The top level is *strategic;* the next is *organizational* and the third is *technology-oriented.* We will look at both organization and technology in the following section.

5. ORGANIZATIONAL PREREQUISITES AND TECHNOLOGICAL SOLUTIONS

At the *organizational* level, the foremost manufacturing companies want to see an increasingly flat structure, squeezing out middle-management levels and emphasizing the professionals. Change in long-established companies is hard, however, because there are too many individuals who resist it.

For restructuring operations in which change was made mandatory by competitive market forces, the question, "Why change?" should first be answered by explaining how the new organizational forms are supposed to work. Though the Japanese are rated as the masters of falt organizational structures, as Figure 1-4 demonstrates, compressing the organizational levels is a job which is not only doable but it has also been successfully executed in Japan.

During the 1950s, and for nearly 40 years, the most forward-looking ideas came from America. Now they are coming from Japan and some of them have not been implemented so far in America and in Europe.

Dr. Peter Drucker suggests that companies organized for survival in the 1990s and well into the twentyfirst century will be federated-type organizations with flat managerial structures. This concept should greatly interest every firm—small or big—as well as its board of directors, top management and organization department.

In a federally organized corporation, the leadership required is one of *ideas* and *knowhow*—not of line of command. Such a company is:

- Rich in professionals, and
- Thin in managers.

In a way, Professor Drucker suggests, the corporation of the 1990s will resemble an orchestra with maybe 300 professionals and only one conductor. An orchestra does not have subconductors and sub-subconductors the way manufacturing companies and financial institutions do today. But it does have a first violinist, and other *top professionals* who distinguish themselves from other professionals by being the *virtuosi,* not the managers.

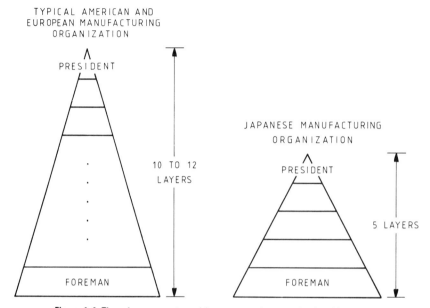

Figure 1-4. There is an urgent need for compressing organizational layers.

One of these *virtuosi* will be the organization expert. Like the other professionals, he or she will have to rely increasingly on influence through *knowhow* conviction, and *leadership*, rather than on formal authority or absolute power.

This is first class advice regarding the functions of an organization department in the 1990s and beyond. The first responsibility in structural studies—and I mean this in a companywide sense—is to realize that the status quo is no longer the best way forward. If we wish to enjoy more of the opportunities and fewer of the risks, we need to understand the changes taking place and to capitalize on them.

George Bernard Shaw once observed that all depends on the *unreasonable man*. When it comes to breaking with the traditional, ossified ways of doing things, the organization department has to be the "unreasonable man"; otherwise it does not perform its duties.

Both top management and the organizational specialists must understand that the 1990s are an age of discontinuous change both in the manufacturing industry and in the technology that it employs. To embrace discontinuous change—which is the organization department's raison d'être—we must completely rethink:

- The ways in which we do things, and
- How we organize ourselves to accomplish our mission.

This is exactly what automation often forces us to do, and we can do it provided we realize what the important components of the underlying technologies are, including their support of world-class manufacturing methodologies. This reference brings into perspective the close knit existing between organizational and technological solutions.

Let's not forget that until quite recently, a fully functional, integrated information system serving the needs of manufacturing was beyond the reach of many companies:

- The applications software was too expensive,
- The required investment in mainframes was too great,
- The cost of implementing, maintaining and supporting the system was inordinate, and
- The concepts on how to do better and to implement faster were missing.

But this is no longer true. Since the late 1980s, a new era began. It is advancing without fanfare but quite surely—though only the leading minority of manufacturing firms do realize the:

- Important competitive implications, and the
- Real cost advantages of implementing the new technologies in order to attain market leadership.

Manufacturing companies no longer have to cope with substandard information systems, and if they do so they put themselves at a significant disadvantage relative to their more advanced competitors. We will see how leadership can be gained in the following section.

6. THE ENVIABLE RECORD OF THE FORERUNNERS

Some companies have achieved an enviable record in factory automation although they are relatively new to this field. Though in a number of cases the changes have been forced on these companies because they made mistakes in their early stages, they have been able to overcome such errors and gain leadership which is most commendable.

Sound organizational principles have played a major role in taking charge of technology. Apple Computers, for example, has divided its Macintosh plant engineering talent into two groups:

- One that evaluates new robots and other forthcoming equipment, and
- Another that constantly monitors and seeks to enhance the plant's present processes.

As part of this search for factory-floor improvements, the plant runs two board assembly lines that use different methods and equipment to produce the same type of printed circuit boards. This lets Apple test different equipment and techniques side by side to evaluate their relative efficiency.

In the Twins headquarters of Nippon Telegraph and Telephone in Tokyo, the company wired each of the nearly identical twin buildings with a different medium: one has optical fibers and the other has a coaxial cable. This permits a real-life evaluation of the efficiency of each system as well as the ability to compare obtained results test their cost/effectiveness, and do the fine-tuning in practically realtime mode.

The enviable record of forerunners starts at the management planning stage. Xerox uses computers to help executives cooperate in planning the company's future. AI-enriched software combines individual forecasts and budgets so that each individual can see how his plans affect the group and vice versa.

Royal Dutch/Shell is experimenting with another interactive technique. Instead of thick books that detail what might happen to the company if oil prices do "this" and interest rates do "that," Shell:

- Built simulators that bring scenarios to life inside the computer, and
- Prompted its executives to create their own models displaying how the firm might react to various events, playing them out on the computer to test their ideas.

In its fundamentals, this approach is not really different from what we do when we implement just-in-time inventory policies, which would have been impossible without computers, simulators, optimizers and expert systems.

Knowledge engineering* today permits building professional expertise into models of the real world in which we work, and experimenting with these models in a meaningful way. What we essentially do is capture the know-how of the most qualified professionals in every firm; they are relatively few and therefore their skills are not available in every position. When this is done, we map this skill into the computer and make it widely available within the organization.

Not only is this a challenging undertaking but it also prompts the establishment of a technology transfer function properly thought out. Slashing inventories and adopting some form of a just-in-time system means employing what the majority of forerunners have already done.

Instead of accepting large quantities of purchased materials weekly or monthly, professionals demand daily or even more frequent deliveries of smaller quantities. Often forgotten in this argument is the fact that JIT works best:

- If the company deals with only a few suppliers, that can quickly respond because of their flexibility or their location,
- If both the company and its chosen few suppliers use simulators and expert systems to fine-tune not only inventory requirements but production planning and quality control as well, and
- If the computer systems of *our* company, its suppliers, as well as its clients are networked and the knowledge engineering constructs supporting an integrative activity communicate online.

Large inventories do not happen by accident; they are the result of long-standing, though not-so-rational, company policies. To make sure that they can meet delivery schedules,

---

* See also D.N. Chorafas *Knowledge Engineering*, Van Nostrand Reinhold, 1990

production supervisors often maintain backup inventories of raw materials, parts and semimanufactured products. Reducing these inventories without the proper organizational and sophisticated computer support raises the chances that the production will miss its schedules and there will be stoppages.

Thus, we come to the conclusion that any manufacturing enterprise that wishes to establish a record of excellence needs a good information system support for reducing costs, controlling inventory, planning material requirements, scheduling production, improving quality and implementing world-class methodologies. This calls not only for computers and communications but also for organization, knowledge engineering and, above all, leadership.

7. A FACTORY OF THE PAST OR OF THE FUTURE?

Leadership sees to it that product planners, marketing people, product designers, manufacturing engineers and industrial engineers closely collaborate to improve quality, compress product development time, or introduce advanced technology. This is not to say that infighting among managers—as well as resistance from the work force—is not surprising when a company embarks on an ambitious, but necessary, improvement program.

The new corporate culture should center on group work emphasizing excellence; but it also observes prerequisites in technology investments. Understanding and automating the reasoning in a complex manufacturing process requires the use of logical models for:

- Handling uncertain reasoning, and
- Describing issues.

Computer programs must be designed for practical use by manufacturing engineers. Under no condition should they use arcane notations or an ardous series of questions and answers. Successful approaches are user-friendly and require expert systems support.

But while expert systems are a key role in the factory of the future, other technology issues are just as vital, if not even more so. Both the corporate and the factory infrastructure should consist of a broadband network:

- Linking all of the plant's design and manufacturing cells,
- Carrying up-to-the-minute information on materials to be stored, retrieved, and transported throughout the plant by automated vehicles,
- Integrating manufacturing cells with the central factory control system, scheduling robots, other machinery and also changeovers, and
- Downloading the programs automatically into the nervous system of the manufacturing plant.

The intelligent information network should verify that the work has been performed to specifications before a workpiece can proceed to the next stage, all of the knowledge and information being stored and transmitted electronically. This is how a paperless factory operates.

Simulators, optimizers and expert systems should see to it that in a fully automatic manner our plant can cut the downtime involved in changing a cell over to the manufacturing of a different product. In its Saturn factory, General Motors has been able to cut the downtime in the changeover of car models from three days to less than 10 minutes—permitting profitable manufacturing at lower-scale volumes.

Saturn is a name that sums up the high-technology ambitions of the world's biggest industrial corporation. As it struggled to come to grips with the Japanese onslaught, GM decided that the Saturn project has to use *start-to-finish* innovation and GM set its course towards obtaining such a result.

After an investment of more than $4 billion and a ground-breaking deal with the United Auto Workers union, the first cars have been introduced to the marketplace in late 1990. But while the technology in use has been impressive, many people miss the point that Saturn does not stress machines or processes as much as it *stresses people.*

"You normally look at advanced technology and think of robotics, automated vision, automatic operations," suggested a senior GM executive. But even in the most automated operation you still have to rely on people. That philosophy is the nucleus of all the technology used by Saturn.

Central to the Saturn experiment was the early decision to develop the project in partnership with all *stakeholders:* The work force and the United Auto Workers trade union, the suppliers and the dealers:

- Saturn workers are all called *team members.*
- The organization is built around action groups and resource groups.

Action groups include primarily the people involved in manufacturing, design, engineering, sales and marketing. *Resource groups* are the support staff in finance, communications, computers, organization development and *people systems*—which is Saturn-speak for the personnel department.

The people systems philosophy is that all teams must be committed to decisions affecting them before changes are put into place. Such changes include, among others, choosing an ad agency and selecting an outside supplier. Workers:

- Are allowed to interview and approve new employees for their teams, the average size of which is ten workers, and
- Are given wide responsibility to decide how to run their own areas.

They are even given budget responsibility. One team in Saturn's final-assembly area voted to reject some proposed pneumatic car-assembly equipment and it went to another supplier to buy electronic gear that its members believed to be safer.

People make things happen. Cooperative systems follow a process of resolving conflicts based on consensus, where participants must be able to *buy in* to 70 percent of the consensus decisions and show a 100 percent commitment to implementing them. This may become the blueprint of the factory of the future as well as of *the management of professionals* as we have seen with the Drucker reference.

8. THE INTEGRATION OF TECHNOLOGICAL INVESTMENTS

At any given level of investment, *our* enterprise would be more competitive if technology and resources were fully and properly integrated. This begins with the people using the manufacturing facilities, tools and equipment and ends with those involved in support functions. All have a role to play in making the integrative system work.

It is precisely in this sense that GM's Saturn claims to be ahead on various technological and organizational fronts. Because of the emphasis placed on the integrative front, it expects the inventory turnover rate to be the best in the world:

- There are 126 receiving docks for components for delivered on the manufacturing floor at the point closest to where they will be used.
- Main aluminium engine and gearbox components are cast on site with the pioneering lost-foam process, Saturn being the first US carmaker to produce automatic and manual gearboxes simultaneously on the same assembly line.
- Another progressive step is the use of waterborne paint (rather than oil-based), which reduces pollution.
- The car body structure is based on a steel space frame with hang-on thermoplastic body panels for all exterior components except for the roof.

Saturn claims to be the first carmaker to use thermoplastic technology and applications on a large scale. The panels can be reused and reprocessed, with the factory reprocessing its own scrap.

In this as in all other applications, to achieve commendable results computer-aided design* must be integrated with computer-aided manufacturing, and I don't mean only drafting work. *If* our blueprints and data sheets are still entered manually into computers at the fabrication and assembly plants, then we do not have an integrated solution. Paper feeds are always time-consuming, error-prone and intensive in terms of costs. From design to production:

- All manual, intermediate steps must be eliminated,
- Design data should be transmitted electronically to factory computers and robots on the factory floor, and
- Programming industrial robots should be done strictly through computer graphics simulation.

GM scientists, for instance, have developed a computer system called RoboTeach, which makes it possible to program robots by simulating them with computer graphics models. The concept is that a technician sitting at a terminal must be able to call up a moving image of

---

* See Chapter 9

the robot, determine the optimal path on the screen, and download the program to a robot on the assembly line.

Intelligent programming approaches can cut the time required to train a robot from about three days to half an hour. This technology becomes increasingly important as the manufacturing industry accelerates its use of robotic supports.*

Beginning several years ago, in the early to mid-1980s and accelerating during the last few years, the realization of advanced solutions has become possible. Driven by development in several key technologies and the introduction of a new generation of software—which means *expert systems in manufacturing*—it is now feasible to provide:

- High performance,
- Fully functional systems,
- Largely integrated aggregates, and
- An unprecedented level of performance-for-price.

There are, however, unique characteristics and challenges associated with such solutions. The careful planning of manufacturing operations seeks to reduce costs steadily by closely knitting component fabrication and assembly and by building the infrastructure necessary to keep a lid on costs. GM plans to cut direct labor to less than 25 hours per car from about 55 hours today by increasing automation, eliminating unnecessary tasks, and consolidating job classifications so that workers can be assigned new tasks on short notice.

Blue-collar workers are not the only ones affected. Companies with the right strategic plans and knowhow in high technology steadily look at eliminating layers of managerial hierarchy—the way it has been underlined in Section 5.

The challenge for all manufacturing firms is to make their plans work at a reasonable capital cost, without requiring armies of technicians to keep the computers and the robots running smoothly. Great imponderables are the ultimate *reliability* of the automated hardware and the *sophistication* of the software we put in place; the action we need to take in both domains is served well through expert systems.

To meet this challenge we need a concept, precise objectives, flexibility in our actions and forethought. Above all we must develop our human capital in being proactive rather than reactive, in leading change rather than opposing it.

---

* Advances in robotics are discussed in Chapter 8

2

Revitalizing Factory Automation

---

1. INTRODUCTION

In a way, it is not difficult to predict the future of factory automation. Ultimately, a major portion of the processing of information that still occupies the routine attention of thousands of people will be accomplished through computers and knowledge engineering. This goes well beyond what we have seen during the last 30 years, as the routine repetitive activities have been executed by rather dumb software.

This is not to say that the more classical data processing chores were without results. Direct results of this second industrial revolution have included:

- An increase in our standard of living, and
- The appearance of new products whose design and production became economical because of automation.

Skeptics would suggest that not all industry has benefited from the data processing evolution. This is particularly true of the service industries: nonmanufacturing productivity, for instance, has risen cumulatively only 1.9 percent over the past 10 years.

Manufacturing companies have faired much better; some have reached 44 percent productivity improvements over the same time frame. Whether in service industries or in manufacturing, companies with substandard results are typically those that devoted far more effort deploying computers than determining how to enhance productivity.

Benefits from the wave of change are realized only by those companies that know how to take advantage of a fast-developing technology. These firms are on their way to realizing the rewards that can arise out of automation tools and techniques.

While the ultimate results are never precisely predictable (though they have to be carefully planned) the rapidity with which we approach these results is materially affected by our decisions, mistakes, and successes. This is just as true of individuals and groups as it is of manufacturing companies:

- If manufacturing firms do their jobs correctly, they will have extensive benefits.
- If they choose improper methods of approach to the many great problems that exist in this field, they will pay the costs but reap no profits.

Today, doing the job correctly involves an increasing number of *knowledge engineering solutions*. Industrial automation systems are fairly to exceedingly complex. Many of the specific techniques and components being used today in manufacturing automation were originally developed in military electronics, but their portability is conditioned by the proper adaptation.

Adaptation is necessary not only with regard to the nature and complexity of the equipment, but also with respect to the type of operations in which it is used. At the bottom, there exists a great similarity between military electronics and industrial automation, but there are also significant differences at the higher implementation layers.

Complex equipment—operated by basically nontechnical people integrates into a very intricate and extensive system of human organization, machines, methods, and procedures. Integration is of primary importance in accomplishing overall operational results. The design of software and hardware is intimately associated with, and affected by, the surrounding complex of men and procedures.

That is why there is something wrong with the policies followed by many companies that often seek piecemeal improvements via islands of automation when they try to reduce the risk that comes with change. Management following this approach forget that in this way they get cornered because of incompatibilities as well as the fact that one part of the system grows faster than the others do, upsetting all balances.

2. A LONG HARD LOOK AT FACTORY AUTOMATION

Nothing short of the proverbial "long hard look" can provide able solutions for the 1990s and beyond. No matter how much money they throw at the problem, factories can never be truly modernized through a series of independent projects, each justifiable on its own terms but lacking coordination with the others.

Such an approach is characterized by tunnel vision. It may work in the short run until eventually a way is found (at great cost) to link these individual islands of automation into an integrated whole; alternatively, many investments have to be thrown away and redone. In fact, experience indicates that the latter is much more likely than the former is, as the piecemeal discrete islands are often inappropriate in the long run.

The long, hard look has, however, its prerequisites. Computer-based automation comes at many levels of sophistication even among computer manufacturers. *Fortune* magazine made this point by quoting Edison D. de Castro, Data General's chairman: "Apple produces almost four times Data General's revenue with about the same number of employees."*

No one component of computer-integrated manufacturing (CIM), is *the* solution. An able solution typically involves many advanced components working in synergy:

- A CAD system,
- A Just-in-time inventory,
- Sophisticated production scheduling,
- A plant-floor data collection,
- Robotics for fabrication,
- Total quality control, or
- Online customer communications.

The desired returns materialize only when all these advances are in place and operate together as a clock. A multidimensional system is shown in Figure 2-1 based on three axes of reference: *resources, activities* and *actors*.

Actors are the society as a whole, the government, the marketplace, the designers, manufacturing engineers, quality controllers, clients, salespeople, and service specialists. Resources are the raw materials, semimanufactured goods, finished goods, plant, machinery, human skills, and knowledge engineering.

The activities are planning, investing, managing, facilitating, standardizing, supplying, developing, distributing, using, and evaluating. In each and every one of these activities, ignorance about the latest developments and a lack of commitment to the cause are unacceptable. Professional solutions have to be *current* and *competent*.

This integrative approach is the No. 1 element distinguishing the revitalized factory automation of the 1990s from that of the 1950s, the 1960s and the 1970s. An almost indefinite list can be provided to illustrate the extent to which organizations tried piecemeal solutions and failed. Such a list will be short if we focus on those organizations that engaged successfully in integrated automation systems.

In Chapter 1 we take GM's Saturn project as an example. Its progressive ideas did not spring up overnight. Many were borrowed from around the world by a team of Saturn workers who traveled more than 3 million kilometers in 1984 and looked into some 160 pioneering enterprises, including Hewlett-Packard, MacDonald's, Volvo, Kawasaki and Nissan.

The main conclusion of this research has been that the most successful companies:

- Provide employees with a sense of ownership,
- Have few and flexible guidelines,
- Impose virtually no job-defining stop rules,
- Plan ahead for system integration rather than patch up afterwards, and
- Focus on quality as a key competitive advantage.

Instead of earning an hourly wage, Saturn workers are paid a salary (the shop-floor average is $34,000), 20 percent of which is at risk. Whether they get that 20 percent

---

* February 12, 1990

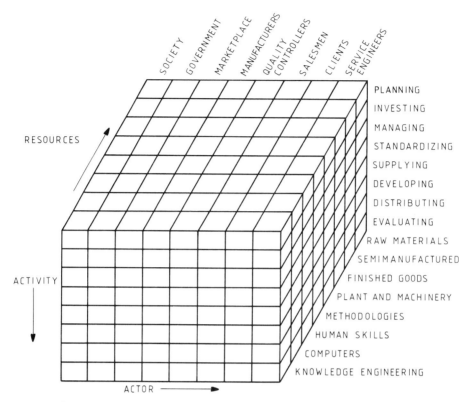

Figure 2-1. Resources, activities and actors can be expressed in a hypercube form.

depends on a complex formula that measures *car quality, worker productivity* and *company profits.*

In the company's first year, employee salaries depend largely on car quality. If a team produces fewer defects than the targeted amount, its members will receive 100 percent of their salary. If they perform even better, they are eligible for a bonus.

People skills are not Saturn's only strong point. Since it outfitted a plant from the ground up, Saturn's team members incorporated into the system an array of new equipment and techniques. Their aim has been to achieve what an MIT study dubbed *lean production,* the Japanese system that uses:

- Half the human effort in the factory,
- Half the manufacturing space,
- Half the investment in tools, and
- Half the engineering hours to develop a new product.

At Saturn, team members rejected the traditional U.S. assembly line. On the so-called *skillet line,* workers ride along on a moving conveyor as they do their jobs. The aim is to help them concentrate on the work they are doing.

Such aims can be reached when we have a plan and know how to go about achieving our goals. The worthiness of such a plan, and its able execution, can take a company to the top of the market. Such a feat will not happen by accident.

3. AN INTEGRATIVE APPROACH TO MANUFACTURING EXCELLENCE

In my work I often hear managements complain that even if they invest in the best technology they experience breakdowns. "The curious feature of the breakdown in technology," the companies often suggest, "is that it is random, uncorrelated between the different applications sectors, and therefore uncontrollable."

The one thing this argument does not consider, and for good reason, is management's own responsibility in the breakdown. The trouble is not that top management and information technology (IT) managers are less able than the other are. The trouble is that their *education* and *experience*—and therefore their *concepts*—are *out of date*.

- Such concepts are built are built for the rather static concepts of the 1960s and not for the dynamic thrust forward of the 1990s.

As Figure 2-2 suggests, an integrative approach will be polyvalent, looking at the customer, the product and the profit center at the same time. Along each one of these three axes of reference, it will evaluate exposure and return.

At the same time, knowledgeable management places a great deal of attention on the *research* necessary to develop the intelligent manufacturing technologies of tomorrow. An example is the Intelligent Manufacturing System (IMS), a project under development in Japan with the Ministry of International Trade and Industry putting up a $1 billion budget.

This plan started after Professor Hiroyuki Yoshikawa of the University of Tokyo suggested to MITI that it was time to make manufacturing hardware and software smart enough so that design changes could be turned with lightning speed into products. Yoshikawa also saw IMS as a way to devise computer standards for electronic links between far-flung factories, thereby boosting productivity.

From manufacturing engineering research to systems integration, companies succeed in their effort if they are willing and able to broaden the scope of their efforts:

- To include a wider variety of technical assignments than used to be considered appropriate in years past, and
- To accept the responsibility for a range of concurrent nontechnical matters that have to be handled.

To a large extent, this has been learned from weapons systems concepts, yet it is not very widely understood today in the civilian industry. The most advanced companies realize it is no accident; giant strides in the application of factory automation techniques have been

26 THE BROADENING PERSPECTIVE OF KNOWLEDGE ENGINEERING

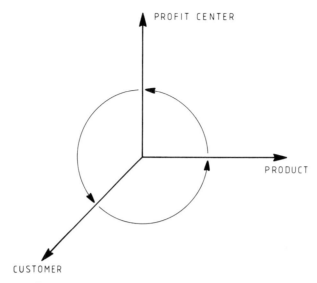

Figure 2-2. Evaluating exposure and return through a three-dimensional approach.

almost invariably associated with projects in which management has assumed a strong position in systems leadership.

The recent history of successful factory automation applications may be summarized in the statement that the key lies in the employment of the proper integrative approach—both in the development work and in the implementation. The basic difference between an improper and a proper solution can be illustrated in a couple of paragraphs.

The *proper approach* is to assign a well-defined *systems responsibility* for the design and integration of the various hardware and software components required in a complete operating automatic factory. Such responsibility should be given to a project leader and a single group of competent, versatile scientists and engineers—the system architects.*

Because of their superior training, these men perceive and conceptualize the solution to new problems. The proper exercise of the systems management responsibility can result in the development of an unusual ability to deal qualitatively, as well as quantitatively, with a variety of nontechnical and technical matters.

A major role within this perspective will be played by *knowledge engineering*. Its able usage rests on professionalism:

- A strong leader will direct his forces with the ruthlessness required to win the integrative battle.
- He will also bet on the winning principle of superior skill and concentration.

---

* See also D.N. Chorafas *System Architecture and System Design,* McGraw-Hill, New York, 1989

This is the only way to tackle advanced factory automation systems that typically employ complex software, robotic equipment and technical methods. No part of a CIM solution will work effectively unless aforementioned ingredients are in place. Building CIM requires:

- A strategic vision,
- Lots of money up front,
- Strong leadership,
- A tremendous amount of patience, and
- The industrialization of expertise.

Along this line of reasoning, the Japanese Ministry of International Trade and Industry (MITI) has instituted a new project known as the *Intelligent Manufacturing System* (IMS).* It uses AI technology to integrate software, hardware and robotics support. The algorithm is:

$$IMS = CIM + AI$$

IMS tackles computer technology as it is projected to be 10 years from now and it thinks about how the technology will be used for manufacturing automation purposes.

United Technologies, Rockwell International and a host of companies from Japan participate in this project. One of the goals is to establish an integrative architecture that will present a standard level of reference in software terms, enhancing program portability the way MIA does. This will permit the user to tackle a wide range of subtle, technical interactions among the various components and subsystems.

One of the major aims sought through such approaches is to integrate into one functional system the contributions of many people and companies each with its own niche of expertise. This makes outside contributions more effective, complementing internal skills and increasing overall performance—as many companies and people are expert in a given domain but lack specific bits and pieces of knowledge other companies may possess.

Unless process design is associated with an analysis that shows it to be compatible with the requirements of the products and the most advanced machines, methods and software for the automated production line will not work that well. Nothing short of a far-reaching systems approach to the entire problem of factory automation can serve to reach effective, lasting solutions, and integrative approaches are key to this reference.

The conclusion is the same in all cases: It is risky to handle black boxes as a portion of a larger, complex system of machinery and procedures. Any advanced system design must be accompanied and governed by a thorough analysis of the entire landscape within which it is to function.

The apparent simplicity of this conclusion should not conceal its very practical implications. For example, this principle asserts that rapid, efficient progress towards modern factory automation will *not* result from the narrow approach of developing robotics equipment, as well as links conveying from one machine to another.

---

* Not to be confused with the DBMS under the same name, which is today obsolete.

4. THE ROLE OF INTELLIGENT COMMUNICATIONS LINKS

Each development project in manufacturing automation requires the establishment of extensive and detailed communication between the basically technical project team and the rather nontechnical operational and supporting people. Experience has shown that such an interaction *can* be successfully established if the communication channels are kept open:

- It used to be that the factory expert on the operational side would have no say about the development of a new automation solution,
- Quite to the contrary, the factory man today must consider himself to be a primary source of input data to the project team, and
- The project team must recognize that as long as the field of automation involves as much operational complexity as technical complexity, the people who do the inventive work must steadily communicate with the production experts.

This emphasis on open communication lines and coordination is made more important as the new factory software and hardware encourage more information-sharing across the company. At the same time, it enables different parts of the manufacturing organization to become more independent of one another—if and when they are properly coordinated.

Experts in the new wave of factory automation appreciate that the more intelligent the software is the more it encourages factories to break up into smaller units of cells dedicated to making specific products. In a properly structured manufacturing operation these cells tend to be:

- Organizationally flat,
- Tightly integrated,
- Self-managing, and
- Highly responsive to evolving market needs.

Expert systems in manufacturing permit a company to become flexible and efficient, thereby reducing the overhead and increasing capacity utilization. The factories emerging from such changes are likely to be smaller but able to generate a greater volume of products.

The new organizational structure, its intelligent software, effective communication links and underlying technology also mean a revolution in relations between *the company, its supplier and its customers*. There may as well be a return to those policies that were dominant 100 or even 200 years ago concerning business partner handling.

For nearly a century, mass production has established an inflexible relationship between:

- Customer, and
- Producer.

Products had to be designed to meet the general but impersonal needs of large markets. The engineering designer, the factory producer, and the salesman became three different

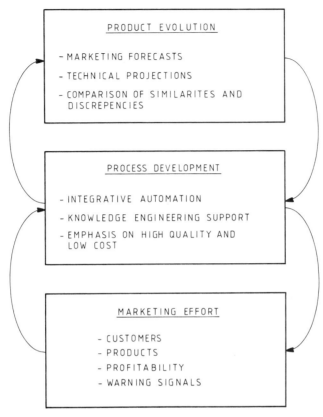

Figure 2-3. A three-way interactivity: product evolution, process development, and marketing effort.

people not necessarily talking to each other, belonging to closed-up parts of the organization.

As a result of this breakdown in communications, customers could not describe what they wanted and manufacturers could not give them options for tailoring the product according to their wishes. The new model runs exactly in the opposite way and thrives on an open environment, as Figure 2-3 suggests. Anything short of open communications is counterproductive.

But with flexible automation, customer service becomes as important as the product. In Tokyo, Toyota has put in its showrooms CAD workstations where the customer, so to speak, can design his car and interactively see it being built-up on the screen as he makes his choices. In Munich, Fiat uses optical disks to show the buyer how the car of his choice behaves on icy roads, on snow, and in the desert in the heat.

As these examples help document, the relationship between the craftsman and his client a couple of centuries ago is being built up again. High technology makes this feasible by "rehumanizing" a link which was dehumanized by mass production:

- The new manufacturing solutions shift the focus from the generalized product back to customers and *personalized* approaches.
- They reestablish close ties between producers, suppliers and end users.
- CAD permits small organizations to compete with skills previously available only in large enterprises.
- The new technologies promote the building of prototypes faster and more economically than ever before.
- Parts and components can be produced efficiently in relatively small batches thanks to computer-assisted manufacturing (CAM).
- From customer wishes to specifications, product images and simulated results, information is communicated rapidly through networks among all parties concerned.

I have chosen to emphasize these points because they are vital for the future of automation. Also, it is not at all clear today that business and industry are aware of their importance or their pertinence to the competitive landscape of the 1990s.

There are many examples of automation in business and industry which underline the employment of methods that might have been acceptable in the past but will be utterly unsuccessful in the future. If obsolete images continue to dominate, then progress in achieving benefits from investments in automation will be markedly slower than it needs to be. Conceptual obsolescence impairs the company's competitive position in a world market that is more fiercely competitive than ever.

Able communications solutions, however, have prerequisites. At the organizational side, they require open lines for information exchange supported without red tape, inhibitions and biases. Or the technical side, they call for any-to-any *intelligent networks*. Such networks are characterized by:

- AI-enriched nodes,
- Broadband lines,
- Quality databases,
- Realtime schedulers,
- Simulators for load balancing, and
- Expert systems for diagnostics.

There exists today in industry a misconception about what an intelligent network is and is not.

There can be no intelligent network to enrich our company's relationship with its customers and its suppliers without knowledge engineering at the nodes, digitization, high quality (low bit error rate) lines, high capacity transmission, nonblocking features, load forecasting and planning, on-line diagnostics and a quality control database. All this amounts to innovation, competitiveness and steady investments - both financial and in human capital.

5. ESTABLISHING A SYSTEM ARCHITECTURE*

A system architecture employs design principles, defines relationships between components and assures the effective interaction between attached devices—whether they be hardware or software. An architecture defines formats that are compatible between dissimilar enterprisewide automation solutions, and it also is a prerequisite for system integration.

The system architect is the person who bridges the gap between the physical and the logical space of manufacturing operations. This calls for paying attention to both the dynamics and the mechanics of the system. There are three prerequisites for establishing the architectural perspectives in an able manner:

- A clearly defined strategic plan elaborated by the board, to be *served* by the technology plan.
- Goals to be reached in products and services, and their cost/effective production and distribution all the way to the market.
- System experts able to work with product designers, manufacturing engineers and quality assurance people, providing knowhow, software and hardware—in that order.

Within this global approach will be elaborated the architectural *definitions,* that is, the grand design. Then comes the need for studying and documenting the concepts as well as the ingenious choices to be made among alternatives.

There are two different ways in which a system team can attack the problem of developing a combination of methods, procedures, knowledge-based software and equipment for the solution of operational issues:

1. In one approach, the development team does a great deal of homework aiming at the employment of the best electronic and mechanical components currently available, attempting to devise ways to combine them in systems that are more advanced than any that have appeared before.

When this kind of invention and design process is carried out with the proper regard for the nontechnical operational objectives of the program, effective results can be obtained.

2. In the alternative approach, the team limits itself to the employment of off-the-shelf devices and subsystems as well as existing production line components, essentially providing the new layout and the interfaces.

---

* See also D.N. Chorafas *"System Architecture and System Design,"* McGraw-Hill, New York, 1989

The building blocks with which the development team works are numerous and elemental in the first instance, and few and complex in the second. In their fundamental form, the analytical and logical problems involved in the design of a suitable system of factory automation look much the same in both cases—what is different is the architecture, which is necessary, and the obtained results.

For instance, interfacing is much heavier in case No. 2, since standardization is not one of the subjects characterizing the computer and communication industry. By contrast, if the architectural study is generic, the results will be much better in case No. 1 since:

- Integration will be more effective,
- The use of incompatible protocols will be avoided, and
- Heterogeneity will be kept to a minimum.

This is not written to mean that Type 2 solutions will not work. As we will see in Section 6 with an an example from General Electric, they can be quite effective and should be preferred when one revamps an existing plan. But a new plant stands more to profit from Type 1 solutions.

Still, whichever choice is made, it is clear that the growth of an aggregate of methods and machines must stem from the large-scale systems study and software development programs that will determine the basic character of the solution sought. The goals to be reached in both cases should be:

- Innovation,
- Knowledge enrichment,
- High quality, and
- Low cost.

These goals are typically sought after by rocket scientists, the members of scientifically trained teams who undertake the assignment of analyzing broad areas of operation in industrial and financial establishments.* Their work helps determine procedures and methods that are most effectively geared to the basic objectives of the business.

If top management comprehends well and early enough the basic facts of the new wave of automation, its decisions will be supported and strengthened by technology. As a result, the company will be able to move rapidly ahead, reaping benefits through cost/effective products.

Yet, while the right policies and a properly elaborated system architecture are cornerstones to successful industrial operations, they are not necessarily enough. Another pillar to greater efficiency is major *investments,* to be done *after* the architecture has been established—not before.

Manufacturing executives at times argue that because the new technologies evolve so rapidly, they should hold off investing in them until the rate of technical progress slows. This is total nonsense; what people who resist the new technology fear is not the accelerating change but:

---

* Using their knowhow from aeronautical engineering and other disciplines, hence the name rocket scientists.

- Its tendency to destabilize the old chain of command.
- Its effect on those skills that are becoming obsolete.

Only the hard facts of life, such as the squeeze on operating margins (Figure 2-4), are sometimes instrumental in convincing management that something has to give: The old methods have to go, or the profitability has to suffer. This practically vinary choice is at the core of the battle to lead a process of change.

The communications perspectives and, therefore, interfunctionality in factory automation, should mean much more than informal cooperation at the lower levels in the organization. It must bring a change in the *structure of command:*

- Middle management layers must be reduced,
- Staffing policies must be refocused, and
- Performance measures must be made to reflect the new realities of industrial life.

The whole command-and-control mentality must be revamped. Central planners should not be making resource allocation decisions at their discretion. The line organization's role now goes beyond operating the established facilities according to targets set in an ivory tower. In an entrepreneurial structure, the line sets the targets and the central planners should be given the leave of absence they deserve.

Figure 2-4. Operating margin of computer and business equipment firms over a 30-year period.

6. A CASE STUDY AT GENERAL ELECTRIC

General Electric's home dishwasher plant in Appliance Park, Louisville, Kentucky, is a good example of what can be achieved by converting the standard mechanized assembly line to automated production. Redesigning the company's line of dishwashers has cut the number of required parts from about 5,600 to only 850—but the effects are even greater at the manufacturing process side:

- Until 1983, when the conversion took place, this GE plant needed 5 to 6 days to produce a dishwasher; thereafter, it required only 18 hours, and
- Overall employee productivity has increased by more than 25 percent, with production capacity at the plant growing by 20 percent.

One of the impressive feats is that such benefits have been possible even though the automation program was installed without shutting down the existing production line operation. Originally, this 850,000-square-foot plant was built in 1953:

- In the late 1970s, the company decided to make the plant its first in what is now shaping up to be a series of automated manufacturing sites.
- The thoroughly revamped operation has become a good demonstrator of what can be achieved through the new wave of factory automation services based on computer-aided design, computer-aided manufacturing, robots, and a rich inventory of software.

Two factors are said to be behind that thorough revamping. Japan and other countries were making serious inroads into the electrical appliance business in America; and GE seems to have seen an analogy with the automobile industry, extrapolating on its long-range implications.

General Electric also saw the opportunity a new departure would present in terms of revamping its company image in the American market. This is an intangible over and above the tangibles of gaining in cost/effectiveness, product quality and overall competitiveness.

The result of the GE decision has been a $40 million program, dubbed Project C, because it was the third in a series of innovations in the company's line of home dishwashers. It was launched in early 1983 after four years of selling the concept to wary employees and union representatives.

The GE plant in Kentucky is a good case study because most of the plant's manufacturing equipment is, to a significant extent, the same as it was before the conversion. But while plastic and metal parts were formerly produced in batch runs and were inventoried, virtually all of them are now made at the point of assembly.

This is one of the organizational changes credited with having reduced the value of the plant's parts inventory from $10 million to about $4 million. Another positive difference comes from the fact that the entire process of:

- Parts production,
- Unit assembly, and
- Warehousing

has been controlled and tracked by computers. Workstations are providing color graphics displays of every machine on the floor while the system automatically locates parts, subassemblies, and other components. Assembly is monitored also by computers enriched with sophisticated software.

One of the applications in optics is the Automatic Camera Recognition System. It aligns the door hinges and the tub structure during fabrication. Since the latching mechanism is an important safety feature, proper alignment is critical, with a permissible variation of only a few thousandths of an inch. At the inspection stage, any unit that has experienced problems during assembly is automatically diverted into a separate repair area.

Project C has not been without its challenges. "If we were starting over again, we would probably spend more time simulating the plant process," said a senior executive. In a similar manner, simulation allows for experimentation on the finer details, emulating how parts and processes flow together so that everything is where it is needed at the right time.

Simulation also makes feasible computer-based tools for training. As the GE and a golden horde of other experiences help document, training programs for a fully automated plant cannot be off-and-on; they must be continuous and permanent—an integral part of the overall operation.

This is an example of a Type 2 architecture approach, as we discussed it in Section 5. It also documents the management's task which requires orderly:

- Assimilation,
- Exploitation, and
- Coordination

of separate source of expertise. Renewal is the key word. With software replacing hands-on skills as well as intermediate management levels, many of the assumptions that worked well in the old manufacturing environment don't work in the new one.

As the example we have seen helps document, the new manufacturing technologies make it necessary to recombine effort and intelligence: They have to become interactive at the micro level; they cannot stay behind watertight departmental walls.

7. REQUIREMENTS FOR A TRANSITION POLICY

Manufacturing companies that mastered the new technology strive to build close horizontal relationships so that product designers work directly with manufacturing engineers, executives from suppliers, production schedulers, quality controllers and marketing people. Effective decisions are reached at the operating level not the ivory tower, and experimentation is taking place on the shop floor as well as in the R+D laboratories.

While this necessarily calls for a new culture, there are also other basic requirements to consider. These are:

- A comprehensive strategy to deal with the evolving markets,
- An exact understanding of the depth and range of *our* resources,
- The wish to develop innovative products and processes, and
- The use of knowledge engineering to enrich human knowhow.

This is in line with the prerequisites we have discussed, namely, the need to focus on qualified professional staff, the implementation of flat organizational structures, and the able use of high technology. Such goals have to be clearly defined by the top management who aim to assure profit stability, strengthen the client base, develop specialized expertise, and steadily improve *technological adequacy* and financial staying power.

But the transition to the new culture cannot be handled overnight. Not only do precise goals have to be established, appropriate for the new manufacturing environment, but the necessary changes must also take place permitting movement towards these goals. Among the basic tools are:

- A *milestone plan,* outlining focal points of transition and their links,
- The development of a *system architecture* which will provide the *metalevel* (higher-up level) for all applications,
- *Experimentation* to define what can be done with the new technologies that could not be done with the old ones,
- The development of *simulators* and *expert systems* permitting such experimentation,
- The *integration* of old and new environments so they can work together,
- The usage of the new *supercomputer,* with the necessary conversion from serial processing concepts to *parallel* processing, and
- *Networks* and *databases* permitting faster, more effective communications and access to stored information and knowhow.

Ably used, the milestone plan and architectural metalevel will help provide a sense of continuity. In October 1988, British Airways chose Boeing over Airbus. Said the company chairman: "The problem with the small Airbus is that it lacks flexibility of size. The Boeing 737 comes in a variety of sizes, from 100 to 150 passengers. Furthermore, we already have a number of Boeing 737 hence it can economize on spares, and pilot training."

British Airways is in the air transport business, but the statement made is directly applicable to a variety of manufacturing plants. Whatever we do, including the decisions we make, must have a sense of *continuity.* This is as true of programming as it is of endusers, databases and computer operations as such.

When we talk of programming aspects in a transition policy, we must be sure to support the psychology of the programmers, leading the team rather than keeping it in the dark. We must also be keen on steadily training our programmers to adapt to the new environment; otherwise, knowledge engineering approaches will not be used.

Considering the difficulties we have to face in the functioning of the program volume, age, and tools to work with, we must be assured that the conversion does not bring interruptions, incompatibilities and delays in other projects. If software packages are chosen, we should add value to them through AI, but we should not change their basic instructions.

As a matter of principle, we should place particular emphasis on realtime and whatever we do must fit within an interactive system architecture. The same is true about enduser functionality.

In any technology-intense environment, the importance of the enduser should be steadily emphasized. We should provide for user-friendly, agile and forgiving man-machine interfaces (human windows); enrich our workstation through AI; and assure fully compatible presentation screens to the endusers.

8. LEARNING BY EXAMPLE AND JUST-IN-TIME INVENTORIES

The predominant concept which came into perspective during the mid-to-late 1980s has been that of growing areas of automation in manufacturing operations that someday will be interconnected by advanced software and communications techniques, as practical technology becomes available.

Revitalizing factory automation and the manufacturing process at large is much more effective, as well as rewarding, when we learn by example. One of the best references substantiating this argument is the implementation of just-in-time inventory management.

Just-in-time delivery and the associated inventory tools assure that suppliers deliver quality-guaranteed parts on very short lead times. But although the sentiments about JIT are pretty similar from company to company, the mechanisms of supplier/customer partnerships vary across the lot:

- Sometimes, just-in-time delivery solutions entail sharing confidential sales projections with suppliers so that production scheduling can be planned further.
- In other cases, they mean bringing in suppliers when a product is being designed, to learn how the parts the suppliers make can best be used.

Often, JIT involves training a supplier's staff in quality control or in manufacturing techniques. Colgate-Palmolive Co. has even let a couple of bottle suppliers set up production lines in new Colgate plants:

- The bottlers get the space rent free.
- Colgate does not take possession of the bottles (and need not pay for them) until they have moved the few feet from the supplier's line to its filling stations.

Digital Equipment has cut some of its lead times for parts in half by programming its computers to transmit sales and production schedules automatically to suppliers. The suppliers view the forecasts as tantamount to purchasing orders for parts; and there is in the works on an Electronic Document Interchange (EDI) system that will automatically issue payment to the suppliers.

Quite often, increased material costs is the background for this type of software. Just-in-time inventory, in essence, took the slack out of the production process, but it also significantly influenced the latter:

- Instead of lending itself to larger standardized production runs, JIT actually focuses more on flexible processes, calling for increased tooling in knowledge engineering to handle such turnarounds.

The traditional method of standardized parts production involves long runs to generate the necessary economies of scale, with machine changeovers tending to be costly because they involve so much set-up time and lost production (running time):

- Since the late 1980s, however, long production runs have been thoroughly questioned, because they tend both to accumulate unnecessary inventory and to disguise slowly deteriorating production quality.

The JIT approach requires smaller runs and this is being successfully accomplished as the latter can now be profitably handled: Computer-based automated equipment allow machine changeovers to be made in a matter of minutes; earlier scheduled changeovers involved hours.

This talks volumes about the flexibility we are after in the new wave of factory automation, as well as what we can accomplish with the tools at our disposal. It also underlines some of the prerequisites. The hardware approach of monitoring a distibuted system is no more the way to go. Instead, our solution should provide for expert systems support able to assure:

1. *Concurrency.* Distributed manufacturing automation has many processes running concurrently. Conventional handling techniques suitable for one process at a time are inadequate systemwide.
2. *Advanced software.* JIT, robotics and other processes cannot be handled through batch-type approaches. Not only must they run real time but they must also benefit from artificial intelligence.
3. *Nondeterminism.* Distributed manufacturing systems are inherently probabilistic. It is therefore difficult to handle them algorithmically; heuristic approaches are more suitable.
4. *Interference.* Monitoring a distributed system alters its behavior. The problem is more serious in a robotics factory, and industrial engineers should provide solutions that get around this constraint.
5. *Communication.* Communications delay among nodes in a distributed system makes it difficult to verify the process state at a given time. Approaches should rest on a realtime architecture.
6. *Interfacing.* User interaction with the monitoring tools can be complex. Expert systems help in this domain by supporting friendly and agile human windows.

Today we can talk with more confidence about improving the outlook of factory automation because the 1980s saw dramatic and significant advances in distributed data bases;* data communication hardware and software; larger, faster, and less expensive computers; sophisticated microprocessors; and supercomputers and robotics. The new wave of factory automation would not have been possible without such advances.

---

* Though in many cases this led to heterogeneous databases, incompatibility in data structures and DBMS would have been just as great if everything remained centralized. Solutions to be provided to heterogeneous databases is one of the most challenging problems of the 1990s.

The concepts and the resources I have just outlined are integral parts of an overall *master plan*. Such concepts, however, did not come overnight and to gain insight as well as foresight we should carefully analyze the successes and failures of early pioneers: we should learn by example.

Over the years, manufacturing companies that failed to do so have been caught up in a wave of dramatic advances in technology to which they do not know how to respond. That is why we must begin to look in Chapter 3 at what has been done through the implementation of *expert systems in manufacturing*.

3

Knowledge Engineering In The Manufacturing Industry

1. INTRODUCTION

Information technology has been used in practically all aspects of manufacturing and it has revolutionized the industry. This is particularly true of recent developments in expert systems, vision systems and natural language, which help accelerate this rapid evolution, becoming the driving motors of competitiveness.

The growing competition in the marketplace sees to it that managers in the manufacturing industry pay increased attention to knowledge engineering because it is expected to have far-reaching consequences on today's design, production and delivery practices. Thus, questions are being asked and issues are being raised regarding developments in AI and knowledge-based systems at large:

- How will these developments affect *our* way of doing business?
- Do they represent real and significant breakthroughs?
- Are they offering new opportunities?
- What is the best way to learn about this new technology so that informed decisions can be made?
- What applications are possible in a manufacturing environment?
- What are the development and maintenance costs of expert systems?
- What are the potential benefits and payoffs?

Parallel to the development of expert systems, significant attention is being paid to *simulation* studies, particularly in those situations that demand stringent process control capability. For instance, the variation in time between steps directly affects process yields and thus suggests the wisdom of experimentation.

Sequential processes performed minutes apart may produce significantly different results from the identical processes performed hours apart because the properties of materials change over time. The slope of a yield learning curve is also a function of turnaround time. Slow feedback delays problem recognition and the verification of solutions.

As an example, semiconductor fabrication is highly sensitive to turnaround time because:

- Definitive functional results are not available until circuits are completely made on the wafer, and
- This happens typically hundreds of process steps after the raw silicon wafers are released into the manufacturing line.

The amount of work-in-process, that is, line loading, affects both turnaround time and throughput performance.

In response to these challenges, mathematical analyses help to generate minute schedules assuring that resources never starve for work or delays are kept at acceptable levels. Queueing theory produces curves that show optimal line-loading levels and other issues helping to resolve inherent conflicts in manufacturing decisions when attempting to maximize throughput and minimize turnaround time.

Planning and scheduling problems are treated in a factual and documented manner in Chapters 11 and 12. The object of the following sections is to help managers and manufacturing engineers find their own answers to the questions asked and the problems outlined in the preceding paragraphs. This chapter surveys significant technological advances relevant to manufacturing applications in the expert systems domain.

Such discussion brings to mind the existing and potential applications of artificial intelligence in the mechanical, electrical, and electronics industries. It gives specifications of the tools required by advanced applications and it outlines the requirements as well as the benefits. It also identifies the dominant technological and implementation trends in manufacturing.

2. THE COMPETITOR'S IMPACT ON OUR OPERATIONS

If our business is at all typical, there is one key factor that concerns us deeply, one factor we think we are familiar with but cannot always forecast: The impact of competition. Technology is one of the tools that can help us better understand just how the forces of competition are likely to affect both our business and our opportunities; at the same time, it is the means by which we can counter the competition.

Firms do not usually enter an industry unless they expect to make money. Similarly, if management does not see much chance to earn profits, it is unlikely to make an investment. That is why it is important to have tools that can help us understand the nature of competition in our industry and probe the limits of our ability to influence the market.

Since we cannot precisely predict how a specific competitor will react, we have to use the best technology available to keep ahead of the game. We do know that the market in the 1990s will reward the low cost producers:

- Are we well below or above that average cost in our line of industry?
- Can we explore why our long-run prices might deviate from the competitors' prices?
- Can we estimate which of our competitors are high-cost or low-cost producers and by how much they deviate from the norm?
- Can we compare our competitors' investment criteria against our own?

The answers to these queries can be explored through financial simulators and expert systems. Figure 3-1 presents an experimental model built to assist management in the evaluation of profitability and budgetary objectives. Three of the seven modules are expert systems; the others are simulators.

Within the *cost of doing business* module, costs vary by type of product and by process. To study and control costs we need detail. A mathematical model enables us to examine each cost class: direct labor and direct material in production, investments, sales and administrative expenses. R+D is the main part of the cost of staying in business.

Once the basic simulator has been done through expert systems assistance, question-and-answer sessions can be framed in a way that most business people can readily understand. Knowledge engineering solutions can take information and produce a very helpful analysis of the decisions to be made.

Visualization shows the results of the analyses preferably in graphical form, assisting the user of the expert system to develop informed opinions. Also, in terms of experimentation

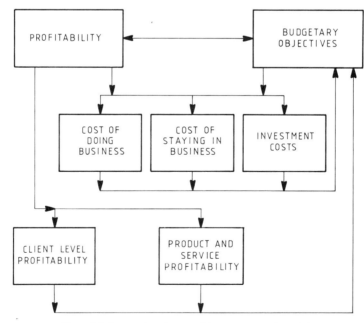

Figure 3-1. A procedure for exercising management control.

the user tries out his ideas until he finds the combination of decision factors that makes the most sense to him. The user may, for example:

- Create a new scenario with a revised price forecast,
- Examine his margin for error, and
- See what the trade-off between value-added and product cost really is.

He may project on cost benefit under different economic outlooks, vary the level of investments, reach different databases to enrich the information available to him, as well as experiment on different alternatives assisted by the rules that have been embedded in the expert system's knowledgebank.

The message the foregoing paragraphs bring to the reader is that expert systems are the industrialization of knowhow and simulation is the way to map the real world into the machine. Expert systems are:

- The first real use of computers for chores that do not involve accounting machines, and
- One of the first practical implementations of artificial intelligence.

One of the contributions knowledge engineering brings to the manufacturing industry is the enrichment of available approaches to designing better manufacturing and more effective product marketability.

The historical separation of engineering from manufacturing has resulted in designs that may not be possible to manufacture at low cost given the available technology. To improve upon this score, knowledge-based systems are being constructed that capture manufacturing knowledge and use it to:

- Identify manufacturing problems while the product is still in design, and
- Recommend alterations to benefit from a synergy of technologies.

Another fruitful implementation domain is the factory floor itself. This reference includes both the design of new manufacturing facilities and their scheduling, as we briefly saw in the introduction.

To optimize investments in new manufacturing facilities, the existing manufacturing base must be evaluated to determine whether it satisfies production requirements, quality objectives, customer delivery demands and corporate cost goals. Knowledge engineering helps:

- Reduce the time required to construct, execute, and evaluate manufacturing models, and
- Provide embedded expertise to automate much of the analysis.

Alternative manufacturing scenarios are thus created and evaluated with the model automatically identifying inefficiencies and bottlenecks. This significantly helps production management.

Production planning can be greatly assisted through expert systems because the process of planning is one of the oldest lines of investigation in AI. Today there is a good inventory of deployed planning systems, as there is a significant number of models that help in research laboratories.

Other applications domains in manufacturing are dispatching and supervising flexible fabrication systems. A knowledge-based construct in dispatching can control work in progress within an assembly factory, including items in storage carousels, robots, conveyors, and workstations.

A knowledge-engineering-based dispatcher maintains a database of products, assembly processes, and manufacturing operations that can be performed at each workstation. As such, it provides a proactive shop floor scheduling and controlling instrument, and it:

- Uses constraint-directed reasoning and heuristic search,
- Provides both long-term and short-term predictive production plans,
- Assures feedback and correction-of-schedule deviations based on realtime information.

Other valuable contributions in the manufacturing industry are made by machine-diagnosis expert systems that construct profiles based on failure modes steadily investigated online. Some models switch from one control plan to another, with the online diagnostic task brings results which perit timely corrective action.

Once a diagnosis is obtained, the expert system makes recommendations to repair the problem. It also constructs a quality database that permits experimentation on future problems, identifying the best way to face them.

These are not projections of future possibilities but references to expert systems already operating in the manufacturing domain. Our competitors are using them and if we fall behind it will be very difficult to catch up.

3. THE EXPANDING DOMAIN OF EXPERT SYSTEMS IN MANUFACTURING

The most advanced manufacturing companies today aim at going beyond the current expert system technology which supports AI constructs in small, well-constrained domains. This is often done by means of aggregation. A group of domain-wise limited expert systems can be connected to work together in a cooperative mode to solve larger problems requiring multiple expertise.

At the same time, capitalizing on the power of supercomputers, the trend in AI is towards more general reasoning. This is possible if the inference engine is fast enough to execute:

- The longer chains of reasoning involved in expert systems aggregates,
- More complex algorithms and heuristics featured by the more recent models, and
- Rather extensive searches needed to reach a conclusion in time for the answer to be of use.

Today, leading-edge organizations increasingly bet on knowledge management, looking at it as *competitiveness* which will manifest itself in the marketplace. This requires advanced research, but the *frontiers of knowledge* cannot be extended except by those who have absorbed the knowledge currently available.

In the meaning of this paragraph lies the real challenge. For this reason, leadership in artificial intelligence increasingly aims at a dual goal:

1. The acquisition of more knowledge through research, and
2. The industrialization of knowhow to make it available throughout the places *our* company operates.

Knowledge management rests on simple foundations: completeness, power, simplicity, and integration. The *knowledgebank* that we construct must be comprehensive. The rules and methodologies that it contains should complement but never contradict one another.

Looking at expert systems in manufacturing as a large purposeful domain, we can appreciate that their implementation focuses along seven lines of reference:

1. The modernization of our technical organization, from CAD/CAM to robotics.
2. The conversion of all active drawings into CAD with the assistance of AI, as well as the use of expert systems in connection to new designs.
3. The production floor, briefly discussed in the preceding section, including production planning and control.
4. Inventory management and sales/inventory coordination (just-in-time inventories).
5. The marketing and sales effort with its customer impact and associated activities.
6. A wide range of maintenance operations—from diagnostics to hands-on product sustainability.
7. Administrative operations including personnel, finance, investments, profit and loss evaluations, planning, controlling and other functions.

Applied to the specific manufacturing domain for which they have been designed, the expert systems that we use should form a *network*. Knowledge management should assure a *complete* development and implementation environment—even if we choose a module by module introduction. This is written because the most striking aspect of artificial intelligence work does not lie in any breathtaking announcement but rather in a methodological approach.

Unlike idealized situations, real manufacturing problems require a complete solution—whether we talk of isolated instances or of a broader range of applications. Real business problems also call for *powerful* but *simple* approaches. Trivia have no place in this connection and complexity has no role in the manager's daily work. When complexity reaches the managerial level, it paralyzes the organization.

These are the key themes of a successful implementation of artificial intelligence in the manufacturing industry. As such, they underline the basis on which expertise is built. Knowledge processing allows us to handle:

- New, tougher problems, and
- Issues too costly for conventional programming techniques.

Besides, knowledge-based technology is now affordable. The point has not been lost on key manufacturing concerns. In 1990 one of the better known computer vendors had 50 expert systems running and management stated that this number is projected to double every year in the foreseeable future.

Another industry, a leading chemicals firm, had 200 expert systems operating in 1990. Management foresees about 2,000 expert systems by 1995, aimed at providing intensive planning and control capabilities as well as a friendly system environment for the operating personnel.

Such interest in the implementation of AI in manufacturing is no random event. It is a reflection of management's conscience that as the competitive nature of the business gets tougher, the surviving companies will be those able to:

- Attain the highest degree of automation consistent with properly planned production yields, and
- Attract enough investment capital to stay the course.

A lot of money will be needed and payback may be slow because revenues must be invested in successive generations of equipment rather than paid out as profits. Or, alternatively, more competitive results must be obtained and in this AI is an instrumental tool.

4. A GOLDEN HORDE OF IMPLEMENTATION EXAMPLES

There are many aspects of the manufacturing industry that make it particularly fruitful for the development of expert systems. The everyday issues of the production floor involve professional judgments that can be expressed in rule form. While design engineering calls for a great deal of ingenuity, here too are several issues that can be assisted through AI.

The implementation of artificial intelligence in engineering and manufacturing focuses particularly on:

- The suppression of loopholes,
- Quicker turnaround,
- Increased design efficiency,
- A more precise match of components and functions,
- Better integration systemwide,
- A quality orientation,
- Improved fine-tuning capability, and
- A readiness to challenge the increasing complexity through simplification.

In terms of management-type assistance, goals sought after with expert systems include a better understanding; "What if" and "Why so" analyses with justification; a test of assumptions; a better conception of processes; an improved working environment; the integration of planning, implementation and control; and as enhanced performance of existing information systems. Another goal involves the provision of an objective basis for the evaluation of different vendor wares.

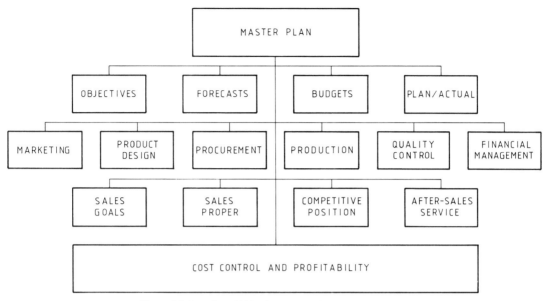

Figure 3-2. Domains of AI implementation in corporate planning.

Many manufacturing companies have paid particular attention to the assistance expert systems can provide to the corporate plan and its ramifications as shown in Figure 3-2. These may be strategic and tactical; they may relate to plans or be operative; and they may be corporatewide or divisional. They have mainly a current or largely a future impact.

If we list the results that have been obtained so far in the manufacturing industry, we will see that in the domain of production and distribution successful expert system applications have focused on:

- Material requirements planning,
- Bills of materials,
- Inventory control,
- Master scheduling,
- Operations routings,
- Shop floor management,
- Capacity requirements planning,
- Lot tracking,
- Job costing, and
- Standard product costing.

In marketing, expert systems have been instrumental in sales order processing, as a support for the configuration of products, and in sales quotations and other issues including sales analyses.

The sophistication of AI constructs varies widely. Among the simpler, earlier expert systems, BACON addressed itself to room planning and layout; FIES analyzed fault isolation and detection; EMES managed the power allocation of electrical components in spacecraft; LUNAR studied the Apollo-II lunar rock sample; PLIDIS controlled the emission of dirty water; POL loaded ships.

SYSAN is an expert system written to help in the documentation of a program library; the ART OF NEGOTIATION simulates a negotiating session, helping to frame a stretagy; COMMUNICATION EDGE assists in preparing business meetings; BLAH gives income tax advice; INDUCE is a structural learning program with good noise immunity; ITA automates robotic inspection tasks.

SCCES is an expert system calculates the risk of various factors involved in stress-corrosion-cracking (SCC), such as crack initiation, when evidence is supplied by the user. The evidence concerns such things as the temperature of the fluid and whether there are crevices in the metal surface. The expert system consists of a knowledge base of rules and facts specific to the SCC domain, linked to plausible-inference software.

DEFT is a diagnostic expert system written by IBM to help diagnose problems during the final testing of large disk drives. Its initial cost has been about $100,000; the payoff is said to be about $12 million in annual savings.

RIC is an AI construct for remote interactive communications by Xerox. It reads data from a copier's internal instruments, senses impeded failure, warns customers and informs repairmen.

GADS, written by United Airlines, stands for a gate assignment and display system. It helps prevent delays that occur when weather and scheduling problems scramble gate assignments for incoming planes.

The diversity of these expert system applications helps document the potential that exists in this field. During the 1980s, and even today, much has been said and written regarding MYCIN and DENDRAL, and relatively little attention has been devoted to discussing the broad range of successful expert applications to real-life problems. We will see several practical examples in this chapter.

5. ACE AND DELTA

Developed by AT&T, Automated Cable Expertise (ACE) is an AI-enriched software that contains distilled knowledge in the form of IF . . . THEN . . . ELSE rules. This knowledge comes from the people who know cables best: telephone company cable maintenance experts. ACE differs from other earlier expert systems in two ways:

1. It manipulates massive amounts of data, and
2. It obtains this information automatically from the instruments in the network and from distributed databases.

This expert system was not developed last night. It has been working since the mid-1980s, assisting the cable maintenance force of the Southwestern Bell Telephone Company. It monitors and analyzes the performance of cable aggregates making recommendations and serving over half-a-million customers in several metropolitan areas.

In its original release, ACE had about 500 rules to follow and, like other expert systems, it was not programmed with all the logical answers to all the possible problems. In terms of implementation, it runs overnight through the cable records of a city the size of Fort Worth, a job which would take a human expert up to a week. In performing its duties, it:

- Collects its information from other computer programs,
- Requests additional data,
- Tests this data against its expert-derived rules,
- Detects recurring patterns, and
- Isolates problems much earlier than a human counterpart would.

The AI construct provides information both on specific trouble types and on locations. When it has a recommendation to make, it communicates it interactively; its main goal is to focus on systems functioning, freeing human experts to work on the causes of problems.

DELTA (Diesel-Electric Locomotive Troubleshooting Aid) is an expert system that assist maintenance personnel in isolating and repairing trains with diesel engine problems. It was developed by General Electric at its Research and Development Center in Schenectady, NY.

In its initial release, DELTA used over 500 rules, the majority of which were aimed at the technician whose job is to repair problems with diesel engines. As a consultation system, it incorporates a software architecture consisting of:

- A domain independent inference engine, but
- A domain specific knowledgebank.

This AI construct uses a hybrid inference mechanism that allows both *forward* reasoning, from facts to conclusions, and *backward* reasoning, to confirm or disprove hypotheses.

Diagnostic and repairing knowledge has been contained in approximately 300 rules; the other 200 rules sustained a help system to answer user queries and to provide additional information. Quite significantly, DELTA can use CAD files stored in Tektronix standard format to help the user further. For example, it shows schematics of the diesel engine or its parts that are being scrutinized.

In a similar vein, the leading automobile manufacturers have equipped their dealerships with expert systems running on the networks they installed in the early 1980s:

- The reason for the communication networks was to handle orders as well as to locate stock (not only dealer-to-center but also dealer-to-dealer).
- The focal point of the networked expert systems today is auto maintenance, but these systems are increasingly filtering into the sales sector that concerns itself with optical disks.

Networked AI constructs are increasingly incorporated into the automaker's production and distribution system so that the dealerships can know fairly precisely when commitments being made will be delivered.

The implementation of such solutions calls for hybrid approaches involving both DP and AI and valid networking and databasing capabilities. ACE and DELTA are good examples in this sense since they demonstrate the integrative issues needed for an implementor to be continuously successful.

While just-in-time inventories, flexible manufacturing systems (FMS), and total quality management (TQM) are core issues, the same can be said of a host of other concepts, with networking and diagnostics playing pivotal roles. Solutions have to be polyvalent; no critical aspect should be neglected.

6. CONTRASTS AND SIMILARITIES IN EXPERT SYSTEMS APPLICATIONS

The growing list of expert systems applications includes the configuration of computer components, a wide range of fault diagnosis, structural analysis, plant operations, automatic programming aids, and computer-aided instruction. If all applications in manufacturing are sorted out along three classes (leaving aside minor implementation areas), then the statistics will tend towards the percentages shown in Table 3-1 which is based on a sample.

In aerospace applications, for example, a space shuttle radar tracking system controls station selection, its output focusing on data inconsistencies. This is fairly similar to an application in the military domain, where interactive AI constructs have been developed for radar signals analysis to assist in interpreting the behavior of electronic environments.

Another expert system written by NASA aids technicians in the diagnosis and repair of the space shuttle's orbiter experimental data handling system. This has a counterpart in civilian aircraft, where an online advisor helps inexperienced users diagnose certain flight plan errors. There is, as well, a fault diagnostic system for electronic test equipment; the expert system:

- Detects problems with communication links,
- Diagnoses difficulties,
- Indicates which device failures could be responsible, and
- Proposes further tests as well as corrective actions based upon its failure hypotheses.

Another AI construct made by NASA maintains the space shuttle's electronic equipment without interrupting its operation. Major airframe manufacturers are now developing similar constructs.

The U.S. Army has an advisory system for combat helicopter pilots. Its mission is to sort out and display incoming sensory, obstacle and threat data to aid in the pilot's decision-making process. Civilian air traffic controllers have at their disposal a monitoring expert system designed to cross-check flight plans filed through uncontrolled airspace and to identify possible safety problems.

The military also has an analytical expert system that identifies optimal attack patterns that hit enemy airfields, selects friendly air bases, and identifies aircraft type and munitions capable of making the attack. It also calculates the effects of the attack.

TABLE 3-1. Application Fields of Expert Systems

Subject:	Percent
Analysis, Consultation, Diagnosis	42.5%
Planning and Design	30.0%
Monitoring and Control	27.5%

In a conceptually similar civilian application, an expert system automates the analysis of data from sensors distributed throughout a system of reservoirs, canals, pumps, and gates. It also helps the control station manage water levels and resources.

In space research applications, there is a knowledge engineering construct providing on-demand advice to technicians performing repairs of malfunctioning modules of the Orbiter Experimenting Program data collection system. A comparable implementation in manufacturing is an expert system that:

- Identifies problem test stands,
- Diagnoses malfunctions, and
- Recommends fixes of automotive transmission test equipment.

In process control, an expert system monitors a nuclear power plant's conditions, diagnoses problems, and provides advice on recommended action enriched with an explanation facility. At a different level of reference, there is in operation a diagnostic and advisory system for electronic circuit board testing; it determines the source of failures and suggests tests for isolating the fault.

There is, as well, a knowledge-based tool in operation that synthesizes a logic-level circuit specification from a high-level functional specification. The input is a standardized representation of a design at the register-transfer level. This representation may also be used as input in a circuit simulation. The expert system produces a logic level design consisting of interconnected:

- Logic gates,
- I/O signals,
- Registers,
- Buses, and
- Transistors.

A functionally correct design is subsequently optimized and sent to a schematic generation package to produce a generic circuit description. This construct has produced many acceptable electronic circuitry designs.

Along a similar implementation line, a simulation model was developed to analyze a semiconductor development line at IBM's Essex Junction, Vermont, facility. The line modeled is responsible for:

- Design verification,
- Process development, and a
- Demonstration of manufacturability.

for IBM's leading-edge semiconductor products. Turnaround time is especially critical in this environment because feedback rates indicate the pace of design and process advances. Throughput is also important to produce hardware in sufficient volume to verify development assumptions.

While expert systems and simulators are complementary processes and should not be confused with each other, they do share common concepts such as *modeling experimentation* and *testing*. In fact, as Figure 3-3 suggests, modeling and testing do overlap to a certain extent, both being based on mathematics.

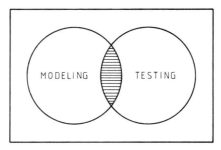

Figure 3-3. The evaluation of system performance rests on two pillars.

Tests should be planned early. What we practically do during a test is to select a typical portion of the system. This can be done by means of:

- Random sampling
- Choosing a representative part, or
- Doing model construction.

AI technology has also assisted laboratory research. An expert system captures the knowledge and techniques of experienced research lab experts who set up the instrumentation for monitoring. It is used both for the training of recently hired set-up personnel, and for new experiments. Another expert system supports international research and the development of automated techniques for a seismological analysis connected to the nuclear test ban treaty verification.

7. PLACING EMPHASIS ON DIAGNOSIS AND FAULT ISOLATION

One of the most promising applications of knowledge engineering is fault isolation. This process poses more general problems than other forms of diagnosis do, in that it is concerned with finding the most likely sources of failure without necessarily establishing in a precise manner what caused the failure.

Fault isolation expert systems have been found particularly valuable when they are used as components of larger more complex aggregates to guide a self-healing functionality in connection with the unit that failed. In that manner, the fault may be:

- Better isolated from the system's operation, and
- More carefully diagnosed by a unit expert.

AI constructs of this type rely on more general reasoning and hence are harder to code efficiently. But new research results have cut ground and have made this implementation more promising.

For instance, in order to assist in computer operations, an expert system advises users and systems analysts to working around data inconsistencies and logical model problems in connection with databases. Another AI construct addresses the fault isolation issue; then it turns directly to diagnostic problem-solving and identifies the following problems as being critical:

- The generation and evaluation of hypotheses,
- The evaluation and selection of relevant sources of information,
- The accumulation and integration of initial findings, and
- The sorting of findings according to hypotheses and goals.

Subsequently, it presents the integration of evidence and suggests just how important a given factor can be to the problem at hand.

SAAB, a Swedish airframe and auto manufacturer, has been using knowledge engineering since 1984. One of the applications is an autonomous missiles system built by using ADA rather than shells. Another AI application is in diagnostics.

Martin Marietta Aerospace has built many successful expert systems. One of them is FIES (Fault Isolation Expert System). It is a rule-based fault detection program designed for space station power subsystems. In its application it has successfully isolated faults, performing or recommending appropriate corrections.

Another expert system, EMES (Energy Management Expert System), manages the allocation of power to various electrical components of a spacecraft. It also permits a graceful degradation of critical system components. Still another implementation is a fault handling expert system for satellite propulsion.

Still another task automation expert system is designed for robotic inspection tasks. It integrates robotics, computer vision, path planning and task scheduling. A similar AI construct built by another engineering firm divides the model into:

- An experimental frame, and
- A model frame.

The experimental frame contains the input data used by the model frame during the simulation and subsequent expert system analysis. The model frame contains the logic used in the simulation and (in a parallel module) the expert system rules. During study, appropriate inputs in the experimental frame are varied to generate various cases.

The pharmaceutical business of Imperial Chemical Industries (ICI) in England has developed a formulation expert system. It is designed to stipulate the correct amount of each ingredient put in medicine:

- If the active ingredient is not very soluble, an experienced formulator would experiment with a highly soluble ingredient.
- This is an established procedure that can easily be translated into rules for expert systems implementation.

The ICI knowledge engineering construct deals both with generic functions and with more complex product formulations. The first release has been a kernel expert system; with time, many more applications may be built around it.

Based on meetings with leading Japanese manufacturing firms, Table 3-2 presents an outlook of knowledge processing in industry. It identifies specific implementation domains,

TABLE 3-2. Practical Use of Knowledge Processing Systems in Industry

Field of Application	1985–1990	1990–1995	1995–2000	2000 Beyond
Design automation		* Software CAD		* New materials design
Factory automation	* Fault diagnosis for equipment		* Automatic operation of facilities	
Office automation	* Scheduling systems	* Intelligent OA	* DSS with model generation capability	
Operation support for large-scale system			* Operation for electric power system	* Operation of nuclear power plant
Consultation		* Energy use consultation	* Total consumption for home	
Computer-aided instruction system	* Course material support, CAI	* CAI adaptive to achievement level	* Training system for operation of power plant	
Language systems				
* Machine translation	* Translation of technical documents	* Word processor with translation capability	* Telephone with translation capability	
* Natural language processing		* Composition machine	* Intelligent word process or * Automatic summary	
Image understanding		* Recognition of freehand drawing	* Recognition in motion	
Voice understanding		* Voice word processor	* Automatic response telephone	
Intelligent robot		* Mobile robot for building maintenance	* Autonomous robot for work in harsh environments	

different phases and tools in the knowledge processing technology, and applications breakthroughs in specific fields.

The basic premise is that of full-scale exploitation of intelligent inference, as well as of creative inference with learning capability. Today's achievements are those of the 1985–1990 time frame, more limited conceptually, but still impressive. Tomorrow's achievements will be those identified by the more advanced entries this table.

In terms of the implementation examples represented in this chapter, the 1990s will be characterized by a dual evolutionary approach. One is the networking of already developed expert systems, enriching them with greater functionality and wider coverage. The other is the growing development of second generation expert systems, based on neural networks and fuzzy engineering, which today constitute the forward looking applications.

4

Practical Examples With Expert Systems

1. INTRODUCTION

In this chapter we will consider a variety of applications to demonstrate the broad field of interest in which AI can be of help. Without presenting a complete list of current and potential expert systems implementation fields, we will discuss both the potential that exists and the benefits derived by many companies through their use of a new type of software.

The justification for implementing expert systems is not found in theoretical niceties but in the results that leading edge manufacturing companies obtained throughout the 1980s when they ventured through the then unfamiliar territory. Their formula has been to:

- Slash production time,
- Improve quality, and
- Enhance customer service.

While the derived benefits are real and growing, it is also proper to be warned. The limited number of presently available knowledge engineers is a factor to consider when one moves AI into real life applications that can benefit from its assistance. Hence, we have to be quite selective.

Generally speaking, the companies that managed technology in an able manner boosted profits and improved their market share. They controlled their inventories better; they planned the time of their people in a more efficient manner, pushing down the amount of manual and clerical work; and they used expert systems to promote worker reeducation. They also exploited AI technology in reducing the amount of *entropy* they had to cope with.

A system's disorder, disorganization, lack of pattern, and randomness of organization is its entropy. Any system tends to increase in entropy over time, but entropy decreases as the level of organization progresses. As Figure 4-1 demonstrates the information environment in any organization ranges:

- From highly structured,
- To highly unstructured

There is a great amount of entropy in the latter, while computational aspects exemplified through AI and simulation tend to bend the entropy curve. The acquisition of knowledge and information signify the power of organization; therefore they act as negative entropy.

Commitment to leadership is a necessary ingredient of any successful effort. This is true at all levels: personal, corporate and national. Knowledge engineering is not magic that will make our problems dispappear. It is a modern tool and the results that we obtain are proportional to:

- Our commitments, and
- Our imagination.

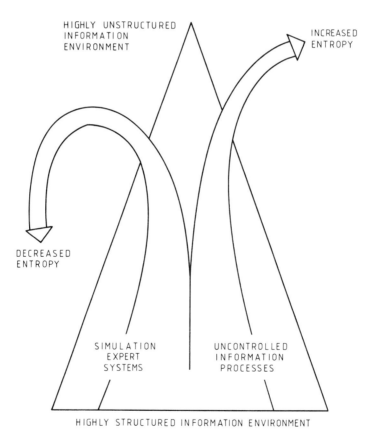

Figure 4-1. Information systems operate in an environment that ranges from highly structured to highly unstructured.

The rapid flow of technology, the improving means of production around the globe, and the clear trend for many companies to gear their strategies and management attitudes to the global market all point to the wisdom of paying attention to ways and means to improve *our* competitiveness. This is not only true of the emerging new, global enterprises, but also of the smaller firms that for survival must move even faster than their big-size competitors.

2. EXPERT SYSTEMS PROJECTS BY BECHTEL CORP.

A Computer-aided Engineering (CAE) project for the construction industry has been done by Bechtel for Hitachi and is operational since early 1989. Its goal is to produce a schedule for assembling a power plant design.

This solution is based upon a simulation program that graphically depicts the assembly of the power plant as directed by the user. Two expert systems have been built into this program:

- A preliminary construction sequencer, and
- A constructibility checker.

The preliminary construction sequencer makes a first cut at sequencing each of the commodity types that are present in the design. Each commodity type is classified and analyzed on the basis of the common practices employed by expert assembly planners. Typically, the planners will sequence large pieces before small ones, pieces farther from the entrance before those that are closer, pieces closer to the ceiling before those that are closer to the floor, and so forth.

The designer uses the individual commodity sequences as a starting place to produce a mixed commodity structure that will eventually be the basis of the construction schedule. At various stages along the way, the constructibility expert system is called to assure that the design is constructible at that point. Among the issues being considered are

- The presence of adequate assembly equipment,
- Possible object interference and other constraints, and
- Fire hazards.

Another Bechtel expert system, the PIMS Advisor, is a prototype developed to demonstrate how AI technology can be used to enhance the process industry modeling. PIMS uses linear programming to optimize models of process plants. It also assists its user in creating and maintaining process models.

A full-featured expert system, the PIMS Advisor has a sophisticated graphics user interface. It is rule based and object oriented, and is implemented with the Nexpert Object knowledge engineering shell.*

---

* See also D.N. Chorafas, *Knowledge Engineering*, Van Nostrand Reinhold, New York, 1990

WELDER is an expert system assistant to a welding engineer who must write welding procedure specifications that conform to regulatory codes. A laboratory test that proves the soundness of the procedure, is required for all procedures used by welders on Bechtel projects. Before a welding specification that is based upon a Procedure Qualification Record (PQR) can be written, the welding engineer must determine if a PQR is available.

WELDER's function is to search the PQR database intelligently for a qualification record that can support the new procedure. When the welding engineer is satisfied that the correct PQR has been found, WELDER automatically writes the welding procedure specification. The benefits of using this expert system include:

- A consistency of logic behind support decisions,
- The avoidance of creating new PQRs unnecessarily, and
- An efficiency in the generation of welding specifications and vendor-procedure verification.

Although WELDER was originally developed for use by a welding engineer, improvements in the user interface had made the construct friendly to maintenance personnel who do not have a welding engineer's specialized knowledge.

Another expert system, the Fan Vibration Advisor, assists an equipment repair technician in determining what is wrong with a malfunctioning fan. The cause of the malfunction is established by analyzing the vibrations:

- After the maintenance personnel answers various questions about the technical details of the fan under consideration, the
- Expert system calculates the vibration frequencies that must be compared to historical data collected by the technician.

Depending on whether or not the spectrum deviates from the norm, the expert system diagnoses the background problem(s) and suggests corrective action. Essentially it operates in a consultation mode, guiding the repairman through his questions in an expedient manner.

Also by Bechtel, a Materials Selection Expert System aids an engineer in making materials selections for petrochemical plants. This project has acquired much of its knowledge from R. White's book on material selection.

A Control Valve Selection Expert System program assists engineers when they develop control valve data sheets or when they review vendor valve data. Control valve selection depends upon many factors besides process data and assimilated calculations. Additional selection factors are:

- Found in various texts, and are
- Based on experience.

This program gathers the sources and experiences, prompting the user with questions able to provide guidance. It also serves as a model for the development of solutions to support other existing design guides.

KATE (Knowledge-based Autonomous Test Engineer) is an expert system designed for NASA, and it is an enhancement of another NASA expert system named LES. The latter is used by engineers and technicians to troubleshoot the liquid oxygen loading system for the space shuttle just prior to launching.

The goal of the KATE project has been to develop an alarm filtering construct using NASA technology to improve operator response during transients. In a more general sense, the utility industry has been addressing this underlying goal for some time. Alarm filtering has been one of many approaches undertaken.

A non-plant-specific expert system has also been developed to assist Bechtel engineering personnel in the performance of safety evaluations. The Safety Evaluation Expert System (SEES) features a knowledgebank that incorporates the safety evaluation guidance contained in the Gaithersburg nuclear discipline safety evaluation training program.

SEES is interactive: The user is required to provide all information needed to address the questions as well as the concerns the expert system raises. Once all queries have been addressed, SEES assures as output a completed safety evaluation, along with a conclusion delineating whether an unreviewed safety question is involved.

3. USE YOUR INTELLECT TO GET RESULTS

The polyvalence demonstrated by these applications sees to it that there is no surprise in the statement: "The newest crew member to board an oil tanker in the Gulf of Alaska is made from silicon." Researchers at Rensselaer Polytechnic Institute have designed an AI construct that helps guide pilots through harbors under a wide range of weather and traffic conditions.

This is a technology transfer expert system as it was originally developed to train pilots in navigating ships into and out of New York Harbor. However, after the Exxon oil spill in Prince William Sound, the Transportation Department's Maritime Administration gave funds to redesign the expert system for realtime navigation in Alaska.

The navigator expert system in reference runs on a microcomputer and is tied into sensors on the bridge as well as to satellite navigation equipment. The latter provide exact readings of a ship's location. The AI construct uses information about fog, drifting ice, or heavy seas to plot out a course:

- Telling the pilot when he has either strayed off course or is going too fast, and
- Giving exact instructions on how to get back on course.

The RAND Corporation has developed an experimental expert system, called TATR, that helps plan attacks on enemy air bases. Knowledge-based simulations were successfully demonstrated by Rand Corporation in the early 1980s; today state of the art vendors offer general-purpose knowledge-based simulation systems.

Bolt Beranek & Newman is working on one that understands natural language questions about enemy positions and then displays the information on a map. Military officials also aim to use AI techniques in training officers and soldiers. This is the case of an expert system that acts as an intelligent manual for the repair of guidance systems by mechanics.

In a different implementation domain, another system enables effective problem solving connected to the diagnosis and repair of performance-tuning problems in telecommunications networks. One of the special features of this approach is the ability to assist in system-to-system communications. Figure 4-2 presents the highlights.

In a similar application, data collected by computerized sensors on military aircraft and satellites are fed directly into AI-enriched computers that alert military analysts. This and similar projects necessitate the simultaneous development of hardware and software, but:

- Since hardware components need to be optimized to achieve the best performance possible,
- The hardware designs continue to evolve even as the software is being written;
- Hence, classical software cannot respond to the challenge. More sophisticated approaches are needed.

This is the case with one of Hughes Aircraft's current projects designing and building custom-made high speed graphics processors for the U.S. Navy. Hughes uses knowledge engineering to write software for hardware that is indeterminate, but the software is ready when the hardware development is completed.

To meet the software challenge, Hughes developers needed a powerful board-level emulator to obtain experimental measurements and predictive data. These were necessary to evaluate hardware and software design concepts for a target system.

The emulator helps interpret and execute the microcode routines as if they were running on a completed graphics processor. Its assistance can be significant in many projects provided that it is:

- Fully functional early in the software development process, to integrate with the other parts of the study,
- Adaptive to new hardware components as these are incorporated during development, and
- Supportive of the hardware and software design process by enabling evaluations of alternative components and their characteristics.

An emulator can be built through classical programming languages, but this is not efficient. In the aforementioned case, Hughes estimated that it would take 1.5 man-years to develop a microcode emulator in that way. Also unsuitable were traditional simulation languages, such as GPSS (General-Purpose Systems Simulator), GASP (General Activity Simulation Program), SLAM (Simulation Language for Alternative Modeling), and SIMULA.

Flexibility and interactivity were other key considerations. The user interface had to be not only agile and friendly but also fully interactive. As for flexibility, it was correctly perceived that it is a cornerstone to success given the aforementioned changing nature of the hardware.

The problem is that once an emulator based on conventional software has been constructed, the target machine's hardware design cannot be altered. If a revised hardware design uses a new integrated circuit, then a new instruction set must be employed and another emulator must be built essentially from scratch.

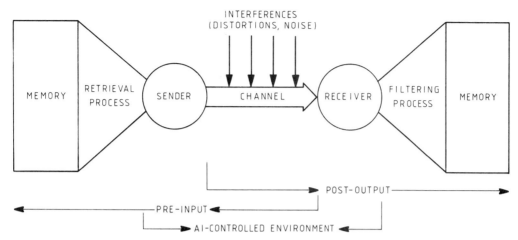

Figure 4-2. The interconnection of information systems is done through the appropriate channel.

What is more, the algorithms of conventional code cannot be untangled and retried at will. By contrast, a knowledge-based architecture permits fast modification. Program development and maintenance under routinely changing conditions is readily accomplished by:

- Specifying new or revised rules in the more classical expert system, or
- Altering the weights in a neural network, in a second generation construct.

A similar conceptual process is valid in the automotive industry since a modern automobile is a computer on wheels with room for passengers and baggage. Virtually every advanced vehicle today is controlled or adjusted by microprocessor, often so automatically that we do not even realize it.

Much of this technology was introduced first in aerospace and aviation applications, but some of the implementation domains are definitely similar. Even more important, the methodology is roughly the same.

For instance, to solve the complex and often conflicting requirements of improved emissions and safety and energy efficiency, today's vehicles require closed-loop, realtime control systems. And these are precisely the technologies that are fundamental to solving aerospace problems through the use of knowledge engineering:

- Aircraft have been fitted with "fly-by-wire" controls for nearly two decades.
- The first automotive "steer-by-wire" systems are just now being tested on prototypes, and are ready for production.

Steer-by-wire is just what it says: Instead of a mechanical steering shaft, the steering wheel turns a sensor connected to a microprocessor. This microprocessor is electronically connected to another one which actually operates the steering mechanism. There is no mechanical connection at all.

Among the advantages are easily designed variable-ratio and variable-assist steering, as well as steering tied into other computerized systems, such as the throttle, antiblock brakes

(ABS), and traction control. AI constructs can help to better performance and to cut costs in each one of these references.

4. EXPERT SYSTEMS PROJECTS BY INFERENCE CORPORATION

As it is to be expected, in implementing knowledge engineering, some organizations, such as NASA, are ahead of others. Among other tasks, expert systems perform fault detection, isolation and recovery for the space shuttle's communications and tracking systems.

Whether manned or unmanned, the missions of NASA are challenging, given the vast web of technical and human problems involved. Reducing the prevailing complexities and their consequent risks is a goal that NASA has consistently pursued enriching it through AI, quite often doing so in collaboration with vendors such as the Inference Corporation.

The Advanced Reasoning Tool (ART), an expert systems shell from Inference, has been NASA's workhorse for years. More recently, NASA used Inference's consultancy services to develop CLIPS—another expert systems shell—and a number of space and other projects using AI-enriched services.

Not all the complexity is in space vehicles and associated structures. Ground control and support facilities are also evolving to become functionally richer, yet the overriding aim is for simpler designs. An example of implementation domains where expert systems have been helpful is the High-Speed Ground Navigation Console (HSGNC) at the Johnson Space Center's Mission Control Center.

The HSGNC is used in monitoring and controlling a Kalman filter during take off and landing. The filter:

- Processes radar tracking data to construct a realtime best estimate of the shuttle's current position, velocity and acceleration, and it
- Generates summary statistics for evaluating the radar station's data integrity and the status of the filter solution.

To improve upon performance of the original Kalman filter version, NASA had to assure both greater reliability and complete control at all times, designing a system that would not overwhelm an operator with a multitude of controls.

The original HSGNC design required three engineers to operate it, as various and contingent events can occur during shuttle ascent and reentry. Efficiently handling the asynchronous data generated by these events is very difficult with procedural programs. In addition, the old version had more than 100 radar-related parameters.

A preliminary study documented that coding the conventional control structures necessary for processing these contingencies would have been a lengthy and complex task. Furthermore, there was the requirement that the new system:

- Offer full human control during operation, and
- Support effective operator training.

This precluded serious consideration of a conventionally designed computer system. Traditional software was also found to be no answer to the goal which was set. By contrast, knowledge engineering provided the solution.

The best design was offered by a data-driven expert system. In this design, the random events themselves triggered the appropriate processing responses and the data-driven approach presented a quick development because there was no need to design, code, and test the complex control structures.

Knowledge engineering as well was the answer to challenges faced by Navistar International (formerly International Harvester Company), the largest producer of medium- and heavy-duty trucks and midrange diesel engines in America:

- Navistar trucks are equipped with various electrical features and components to meet the variable needs of the truck driver.
- The larger electrical components are connected to the battery by a wire harness, composed of wires, terminals, connectors and circuit boards.
- As a result, the wire harness is one of the most complex parts of a truck to design and manufacture because it is composed of many different parts.

Furthermore, the designer must work within strict design constraints to protect its electrical integrity. He must remember that the precise number of wire harnesses on a truck vary in size and complexity. Most trucks, however, have several harnesses that must be quality controlled.

Quality control of the wire harnesses requires the analysis of a huge number of variables, a task that is time consuming and calls for a great deal of expertise. And in an age when only low-cost producers can survive, the costing aspects cannot be forgotten. Typically, a cost estimating expert receives a wire harness design from the engineering designer and evaluates it for cost efficiency.

By evaluating the design, the expert can ascertain how individual variables affect the overall cost of the wire harness. But to analyze the cost efficiency of a wire harness properly, he must consider a large amount of data, including:

- The number of wires and terminals,
- The number of parts and their tooling, and the
- Materials and labor costs required for each one-wire harness.

No wonder that in the past such estimates were not always accurate. Also, since the estimating process was so lengthy and expertise-intensive, the experts were not able to audit and evaluate every wire harness although the competitive nature of the truck market underlined the need for auditing all wire harnesses.

As a result, Navistar's management investigated artificial intelligence technology as a means of saving money as well as to compete more efficiently in the market. They chose Inference's ART as the shell for this application providing the cost estimator with an AI model so that he can be assisted in making recommendations to the designer.

The Navistar and Inference knowledge engineers codified the tooling process, guaranteed the possibility of experimenting with alternative wires and connectors, and reduced the amount of time required to estimate the cost of the wire harness, thereby effectively controlling costs. This also assisted the purchasing department in evaluating outside vendor bids.

Navistar also analyzed the existing constraints that would affect the project's design and implementation. These included:

- The need for a comfortable human interface, and
- The ability for a user to be able to change values within the system.

Navistar wanted the system to be at least partially generic so it could eventually be used as a model, building other cost-estimating approaches. The obtained rule-based solution not only translates human expertise into rules understandable by the computer but it also provides the required flexibility.

5. MULTIMEDIA EXPERT SYSTEMS FOR TOPOLOGICAL APPLICATIONS

The subject of this section is the development of topological information processing systems with fully interactive characteristics. This is becoming increasingly important for planning purposes as well as for facility management, the reference being particularly true for utility companies.

Recent progress in computer and graphics technology facilitates the study of topological solutions, but there are also more challenges to overcome. One of them involves managing multimedia approaches, including:

- Figure data of maps,
- Character and numerical information such as names and attributes of the figure data,
- Text embedded in documents and memos, and
- Images such as landscapes and aerial photographs relating to the figures.

Another challenge involves providing an efficient and intelligent interface for multimedia purposes through user-friendly instruction, preferably in a natural language.

To cope with these problems, Hitachi knowledge engineering developed a new geographical information processing system: the General Topographical Land-Use Expert system (GENTLE). Written in Lisp, it handles a large multimedia database through a relational model and a knowledgebank. Figure 4-3 shows its architecture.

The main reason for using a knowledgebank, the Hitachi researchers suggested, is to increase the efficiency of information retrieval. A proper retrieval procedure is deduced from the knowledgebank to search each relational table effectively, drastically cutting the time for an exhaustive search of the lower layers of the multimedia database.

The knowledgebank usually stores the system structure and database schemata. Therefore, it is the center portion of the general-purpose geographical system. Its role includes:

1. Providing a knowledge-enriched outline of the database used in the system.

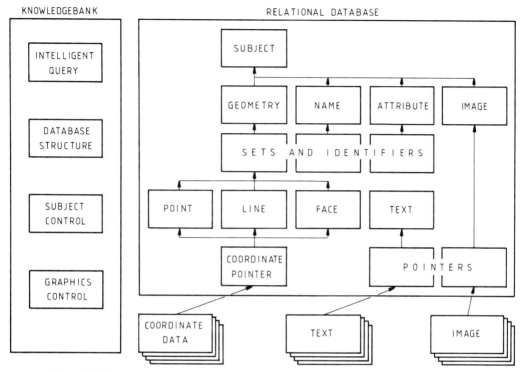

Figure 4-3. There is a correspondence in functionality between a knowledgebank and a relational database.

This is described in the knowledgebank together with the information on its usage. Therefore, the system can be easily implemented with various existing databases without rearranging their contents.

    2. Intelligently surveying database schemata by consulting the knowledgebank.

As the researchers have demonstrated, in a multimedia environment search heuristics can be much more efficient than search algorithms are.

    3. Providing for efficiency in multimedia retrieval by producing a plan based on database schemata.

The efficiency of this solution has been compared to that of traditional approaches using a relational form. The same topographical map data have been employed.

Obtained results help demonstrate that the upkeep and accuracy of information retrieval, and most particularly, the speed of the knowledge-enriched method, is higher by *two orders of magnitude*. The reason lies in the fact that the AI-based solution can effectively utilize the database managers specialized for each medium (text, data, graphics, images), instead of the inefficient one-way relational retrieval process.

In this design, the contents of the knowledgebank are divided into four classes, each one represented by a frame-based hybrid type approach.

1. Query analysis and meaning-extraction frames.

These frames analyze the ambiguity involved in the natural language, extracting the structure and meaning of the sentences. Hence, they clarify the applicability of the complex commands given by Japanese language sentences.

2. Database structure and meaning frames.

These memorize the organization of the individual database, the relative structure of each medium, and the meaning of each relation.

3. Subject map frames.

Such frames record knowledge about such features on the map, as buildings and roads. When an insufficient retrieval query is made, this knowledge is used to infer the missing information.

4. Graphic control frames.

Their goal is to store the graphic knowledge of color selection, line types, hatching patterns, and so on, to be used in the system. When a plural number of subjects, such as buildings and roads, are to be displayed at the same time, this knowledge effectively contrasts these subject classes on the screen.

In the GENTLE solution, the system configuration data and semantic information of the database are also treated as *knowledge*. In particular, a schema of the database is stored as the semantics of the database and it plays a large role in improving the efficiency of retrieval.

The Hitachi researchers have also constructed natural language interfaces to the applications. An interpreter incorporates case-frame instantiation and pattern matching to analyze natural language input, translating natural language user input into an internal representation meaningful to the solution that has been adopted.

As this example helps document, integrated intelligent interfaces are "a must" for the 1990s. Natural language facilities that help in expert consultation systems are being increasingly constructed with work beginning on integrating multiple modes of man-machine communication, such as those provided by:

- Natural language,
- Graphics, and
- Various pointing devices.

Another goal is to produce a text scanning and summarization expert system able to analyze and interpret a high volume of short natural language texts and to classify them into user-defined categories facilitating the extraction of critical information for routing and database handling.

Viewed under this perspective, the Hitachi expert systems application in topological issues is polyvalent and, conceptually, it contributes a great deal to other domains of implementation in the manufacturing industry. We must always be keen on transforming

6. IMPLEMENTATION AREAS THAT HAVE GIVEN COMMENDABLE RESULTS

technology from one field to another and this case study offers a good example of this line of reasoning.

The aim of this chapter has been to present the reader with implementation examples that have not been overexposed to the public eye. Some, like those by Bechtel and Inference, were made on a consultancy basis; others were rule-based programs done by user organizations and performed tasks that could perfectly well have been achieved in other ways, but in a less efficient manner as, for instance, the Hitachi topological application.

As the range of AI applications steadily grows, some kinds of expert systems are inevitably substantially underrepresented in the published literature. Some others are proprietary, and only summary information can be obtained about them. There is, as well, a large number of small constructs that have been built using standard shells for management's familiarization, but they have not been converted into fully operational expert systems, documented or named.

AI constructs treated as commercially confidential by the companies concerned inevitably include some of the largest and most interesting applications. The same is true of expert systems still under development, among them some of the more complex kinds.

Made at the Boeing AI center, a 1,000-rule prototype of a knowledge-based construct aids shop floor personnel who assemble electrical connectors, to select the correct tooling and materials. This prototype has demonstrated:

- A 100 percent accuracy, and
- Significant reductions in process specification search time.

Both experienced workers and new personnel have been pleased with the advice provided by this expert system and for good reason.

The electrical connector assembly for small lot sizes is largely a manual process. To carry out major assembling operations, shop floor personnel consult typically ten, but sometimes up to thirty, process specification documents to determine the allowable tools and materials for this task:

- The relevant documentation is about 200,000 pages, and
- twenty-thousand of these pages pertain to connector assembly.

There are approximately 150 connector types to produce and 200 different precision hand tools to use. Even if shop floor personnel interpret the documents correctly (which is unlikely), the consulting process can take considerable time.*

---

* See also the CASES experience at the IBM Endicott factory in Chapter 6.

Originally, the Boeing prototype in reference covered 25 percent of the work done in the shop, which the company considered adequate for testing the expert system and for determining further requirements. The full-scale production model, was ported however, on larger workstations and the model practically covered most cases.

This example brings into perspective an application representative of a variety of problems faced by factory floor personnel, particularly in industries concerned with small lot batch processing. At the same time, the project shows that the use of AI technology has virtually no limitations. One needs only to understand it and to have the imagination to apply it where bottlenecks are seen in the production process.

A multimedia prototype expert system for diagnosing the causes of cracks in concrete has been developed by the National Institute of Standards and Technology (formerly NBS). Known as CRACKS, this construct includes:

- A database for maintaining structural descriptions, and
- An image base for storing digitized photographs and drawings of cracking.

The expert system can also be used to measure the rate of deterioration of a structure by comparing observations over time.

In England, Project Appraisal Resourcing and Control (PARC) aims to address key problems in project planning, management and completion, through the use of expert systems. The issues involved include:

- Uncertainty in decision processes,
- Skills transfer,
- Learning capabilities,
- Decision making under conflict, and the
- Selection and application of appropriate information elements.

EXMAR, another British expert system project, aims at spreading marketing skills more broadly through a company. It analyzes a range of product and market-related data; assists in the identification of detailed marketing strategies to be followed; and forecasts the results of a given strategy in terms of sales and profits.

Until British Coal made the decision to develop an expert system, the problem of predicting whether a new event would give rise to unusual *methane emissions* had never been properly addressed. The information needed for the prediction is complex, rather specialized and extensive:

- Expertise is needed from geologists and ventilation engineers, and
- Reference must be made to historical data and case histories that indicate where problem areas may arise.

British Coal engineers have, however, been able to produce an expert system that gives broad conclusions useful to mine workers.

Users say that one important feature of the methane expert system is that it is a teaching aid: The enduser gains knowledge about assessing methane risks from the questions the program asks. According to British Coal, this interactivity has been instrumental in improving know-how in a companywide sense and has also provided significant savings.

Provisional estimates indicate that in the case of British Coal about 2 million pounds ($3.6 million) worth of coal production can be lost a year in a single coal board area through unexpected methane emissions. As the company has 10 areas, the potential savings are in excess of 20 million pounds.

After the tests of the developed prototype were concluded, the methane detection expert system was installed in one of the main coal mining areas and is now used by the professional ventilating engineering staff. Results have been so encouraging that British Coal is considering a more comprehensive version that will be adopted for routine use in all mines.

Another expert system project by British Coal is *heatings*. It is intended to become a realtime expert system linked to environmental monitoring in mines. Engineers want to be able to predict and to prevent the hazard of spontaneous combustion caused by the oxidation of coal whenever it is exposed to air.

Such a process produces heat, carbon dioxide and carbon monoxide. If the heat is not quickly removed, the exposed coal face can smoulder or burst into flames. The proper design of roadways and faces, combined with care during mining, reduces the risk of this hazard, and the AI construct can be of instrumental assistance in this process.

Among other developments, the Canadian government has designed a Residence Expert System (RES) for taxation purposes. It determines a legal domicile on the basis of the time a person spends in Canada and it is said to have led to some interesting patterns: Rich people spend half their time in Florida.

The American government, too, is active in developing expert systems for taxation reasons. Some years ago, the Internal Revenue Service (IRS) commissioned Bolt Beranek Newman (BBN) to develop a fairly complex expert system for the automatic auditing of tax returns. Following the success of this project, the IRS sponsored nine other AI constructs. It also sends its people to universities for a year to learn knowledge engineering and specialized expert systems for taxation reasons.

5

AI In Network Design And In Power Engineering

1. INTRODUCTION

The examples that we have seen so far in the first four chapters of this book provide the reader with evidence that, in terms of implementation, expert systems generally fall into two categories: Those intending to replace an expert, and others aiming to assist him.

The second bet is more serious than the first, if for no other reason than because our understanding of the intelligent network designs of the 1990s as well as our handling of complex nuclear engineering problems is such that we cannot expect to create fully automated approaches now. Hence, both fields are fertile grounds for "assistant to" expert system implementations.

One of the imaginative projects that have been made in the communications domain is a scenario system that creates displays, allows modification, helps test alternatives, and saves those solutions it helps to develop. These can be used subsequently as input to a communications network throughput and response time simulation model.

Chapter 4 has underlined that diagnostics and maintenance are fertile terrains for AI; this is just as true of networks. There is, for example, a successful diagnostic expert system operated by NYNEX in 43 maintenance centers throughout New England. NYNEX has released figures for savings of between 4 and 6 million dollars per year. The Maintenance Administrator Expert (MAX) diagnoses problems with residential and small-business telephone service, and it is designed to increase the accuracy of technician dispatches by providing an expert interpretation of the data on the trouble reports it screens.

Another AI project is a knowledge-based design aid for electronic network analysis. Its strong phase is at the design level in which optimal conceptual configuration must often be compromised to accommodate technological limitations. This system:

- Emulates human experts in analyzing waveforms produced by a circuit simulator, and it
- Recommends and implements changes to eliminate problem areas that would impede manufacturing.

Typical design changes include altering wire lengths and adjusting the electrical properties of components. As we will see through specific examples, similar goals have been followed in nuclear engineering implementations.

Another aim followed both in network design and nuclear engineering involves *hybrid systems*, the merging of AI with traditional DP routines. This is particularly true of realtime applications.

Though expert systems were initially seen by many as an entirely new form of software for stand-alone usage different from those of conventional computing, attention is increasingly being focused on forming links between expert systems and the conventional problems of data processing including the interface between AI constructs and database management systems (DBMS).

This reference is as valid regarding the development of intelligent front ends, complex software packages and information retrieval systems, as it is to applications in realtime

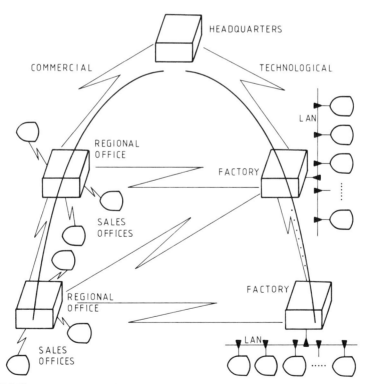

Figure 5-1. There are two networks which have to be integrated: one for marketing, the other for manufacturing.

process control. Realtime hybrid system solutions are of significant interest to business and industry and they will be even more so in the years to come.

Figure 5-1 presents as an example the solution adopted by a manufacturing company that established an any-to-any network interconnecting headquarters with commercial offices and factories. Both the commercial and the technological information systems are enriched with expert systems that communicate on-line with one another.

On the other hand, it is also true that despite the undoubtedly high level of effort put into off-line expert systems worldwide in the last few years, it is proper to observe that there are still very few substantial realtime AI solutions in regular use in industry. The present chapter addresses itself to this class.

2. EXPERT SYSTEMS FOR NETWORK DESIGN AT BOLT BERANEK NEWMAN (BBN)

The design of large, complex networks requires macroengineering concepts properly conceived and executed. Both their conception and their execution can profit from the use of expert systems.

With macroengineering, the problem may be well defined—which means there exists a fundamental knowhow—but it is larger than one entity to solve. This means bringing into the picture many corporations that contribute their parts to the project, the whole process posing severe coordination problems.

Expert systems can be of assistance because solutions to the problems that arise are required simultaneously in many places. This has been in the background of the expert system construct known as DESIGNET.

DESIGNET, an expert system developed by Bolt, Beranek, Newman, is a successful automated data network design. This is one of many products that has come from BBN in the AI domain; others include RS/Expert, a statistical advisory construct; radar intercept (an operator training aid); propulsion system (a teaching tool); pharmaceutial plant configuration; software to control robots (in an NBS automated factory demonstration laboratory); and a natural language interface that translates English queries into computer commands.

When asked why they bet on expert systems for network design instead of staying with traditional approaches, BBN executives underline four main points:

1. Textbook solutions are not very good.

Theories regarding approaches to network design still are where software used to be 25 years ago. This hardly responds to present-day requirements.

2. Devices, protocols, and tariffs are constantly changing.

Protocols are not only incompatible among themselves but there are also in flux. To cope with such change, we must have powerful tools that permit us to be and remain flexible and to respond quickly.

3. Classical design processes take too long.

Not only do they require a number of experts who are in very short supply, but the development time itself is also inordinately long when compared with the requirements of a very competitive market.

    4. As systems get complex, we must assist the human intellect.

The critical problem is the management of partially understood, and poorly defined network systems. Complexity will increase in the years to come and if we do not sharpen our tools we will not be able to face the developing challenges.

To face the fast developing challenges, advanced mathematical and graphical techniques must be implemented. In connection with the BBN expert system for network design, one of the most basic is the topological structure we will examine in the next section:

- An analytical model is invoked to calculate the average packet delay in the network, or along any path.
- Additionally, algorithms for fast connectivity analysis are used to analyze network reliability.

For example, if the network is not triconnected, so that the loss of two packet switches partitions the network, this is automatically detected and the separation pairs are highlighted on the map through visualization.

A mathematical model representing the network routing algorithm assigns flows to paths. Lines and nodes that are very busy are highlighted on the map so that network bottlenecks can be easily identified.

An object-oriented programming language allows the implementor to define *objects* (abstract data structures) that have properties enabling them to be individually manipulated. The network engineer can establish objects for:

- Tariffs,
- Devices,
- Protocol, and
- Algorithms.

Such a feature has two consequences: Modularization is facilitated and the implementation of the system can be made to mirror the expert's model of network organization. The protocol objects are layered much as they are in the ISO/OSI standard.

The knowledgebank of the expert systems contains detailed models with language on devices, software, protocols, and tariffs. In terms of output, the expert system responds in relation to three basic criteria:

1. Switch utilization (the blue line)
2. Link utilization (the red line), and
3. Cost (the green line).

The AI construct keeps track of different solutions and presents them graphically. *Commands* and *processes* are based on designer experience, but *values* and *objects* expand beyond designer limits.

One of the modules, the *Configurer,* has the job of configuring packet-switching networks when the design phase has been satisfactorily completed. It knows all the rules necessary for configuration studies and (like DEC's XCON, which we will examine in chapter 6) it produces as an output a bill of materials.

DESIGNET, used over several years for all the network design work at BBN, has been undergoing major improvements. Some of these developments have extended its capabilities so that it can deal with technologies such as voice transmission, integrated voice/data communications, and SNA networks.

One of the expert system's interesting features is its *critic* capability that enables it to provide a critique of the quality of the network design. This developed into an *advisor* that makes suggestions on how to improve the design.

The BBN management also envisions the development of an automated design capability to handle trunking services which require that topological design takes place in real time in the network operations center. Another offshoot is an expert network performance analysis module. Its mission is to:

- Organize the voluminous data received at the network control center of a packet-switching network,
- Look for trends and patterns in this data, and
- Point out to the expert network analyst potential causes for performance problems.

A version of this network design model has been turned over to the sales engineering department of BBN. Like XSEL, the twin AI construct of XCON, its implementation in the marketing domain is that of generating customer proposals.

3. GLOBAL NETWORK DESIGN THROUGH AN EXPERT SYSTEM

A key part of the network design process is to produce the specific configuration, or *topology*. An optimal configuration would have a minimal cost consistent with the traffic matrix and the other requirements the network must satisfy while considering the component phases of the problem properly.

Network designers find it convenient to divide the topological design for packet-switching communications into four phases:

- Initialization,
- Terminal homing,
- Host homing, and
- Backbone.

In the initialization phase, the data on traffic and other requirements is placed in a mathematical form and is processed by the network design susbsystem. A vital part of this activity is the application of a protocol model to the raw traffic matrix, since much of the traffic carried on a packet switching network is used for protocol overhead.

For instance, despite the fact that the X.25 protocol allows ISO/OSI Level 3 acknowledgements to be piggybacked on data packets, the large majority of host implementations do not take advantage of this facility. The result is that at Level 3 a response/confirmation packet is created for each data packet received. Thus, half the packets on the network may be simple overhead.

Consequently, a network design that does not have a detailed protocol model runs the risk of producing outputs seriously underconfigured for the traffic that must be carried. Equally important configuration questions involve:

- The placement of terminal concentration devices, generally packet assembly/ disassembly (PADs) and multiplexers.
- The choice of terminals to connect to specific concentration devices, and
- The location of terminal access lines.

This problem is one of clustering, wherein the designer trades off the cost of concentration equipment against the cost of circuits. At the same time, he is constrained in solving this problem and he must be careful not to exceed the fanout and throughput of the various access devices as well as the bandwidth of all circuits.

The expert system can offer a significant assistance in this connection as it helps the network designer not only in experimenting with alternative configurations but also in optimizing the one that is selected. It will be recalled from Section 2 that the crucial factors in optimization are:

- Switch utilization,
- Link utilization,
- Investments, and thus cost.

Conceptually, the approach elaborated in the preceding paragraphs can be expressed in a graph as the one shown in Figure 5-2. At the top is the method builder followed by the overall configurator. The former is a metalayer to the latter.

The middle level is characterized by task management with many nodes depending from it. Task management is a metalayer to the nodes that among themselves benefit from common services.

The lower level is composed of special software, a network operating system, and hardware. It is preferable that they be characterized by networkwide portability, which suggests a portability metalayer above them.

The third phase of a network topological design is also a clustering problem in which one determines where packet-switching nodes (PSNs) should be placed and how different devices including PADs should be connected to packet switches. The designer is constrained by fanout and throughput limitations of the network equipment as well as by the capacities of the circuits. Therefore, he can find significant assistance in his study through the expert system.

The challenge of backbone design comes up once the designer has determined where packet switches should be placed. In this phase, the task is one of determining how to connect packet switches to each other. In solving such a problem, the designer is constrained by:

- The throughput of the switching equipment, and
- The capacities of circuits.

The backbone network must be configured so that network availability and reliability

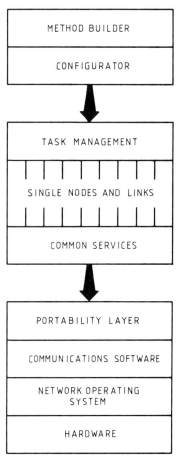

Figure 5-2. Task management and common services have preprocessing and postprocessing requirements.

goals are met, while delays and other performance characteristics are acceptable. To meet these goals, a number of mathematical problems must be solved.

A model must be constructed for the network routing algorithm so that end-to-end flows will be assigned to the same paths through the network that will be used in actual operation. Once the flows are assigned, the utilizations of all switches and circuits can be computed. A mathematical model can then be used based upon the queueing theory to compute the packet delay.

Similarly, algorithms and heuristics are needed to analyze network connectivity and to compute reliability estimates. This is written in the understanding that the design of the backbone is much more difficult than the design of the access area, and this exists for a number of reasons. For instance:

- The access area is constrained to possess a treelike structure with packet switches at the roots of the trees.

- This limits the options available to the network designer and makes the design problem easier.

Additionally, the access area design problem is of a clustering nature and clustering issues have been studied for quite some time. Switch placement problems are quite similar to warehouse placement problems; in both cases we trade transportation costs for warehouse costs much as we trade line costs for switch costs.

Thus it can be seen that in a theoretical sense several phases of topological design occur in a well-established sequence involving PAD placement, node location, and trunk establishment—though, in reality, this process is iterative. For example, the placement of nodes may render some choices for PAD locations suboptimal, or the interswitch circuit costs excessive. Hence, in practice, the network designer moves back and forth among the various phases.

Subsequently, once the topological design is complete, the final phase of the network study involves the detailed configuration of each of the components. The difference between this phase and the topological design is substantial, though this may not be readily apparent.

In topological design, we deal with abstracted, simplified representations of devices. We must understand the costs of components and their performance characteristics, but we need not know about racks, cabling and power as this level of detail is not necessary to solve the optimization problem. But once the overall conceptual network design is completed, we must produce a detailed list of the equipment that will be placed in the network, and plan for its interoperability.

Most importantly, the network design problem is made challenging by the volume of information that must be comprehended by the designer:

- In a very large network design, we may be dealing with thousands of attached host computers of all types and sizes and hundreds of packet switches.
- As a result, the traffic matrix may very well have several million individual entries that behave in a dynamic manner.

Since we cannot rely upon pure algorithmic solutions to this problem, we must depend a great deal upon the intuition and experience of the systems architect. Faced with complex network designs and growing user requirements, system architects and designers can draw significant assistance from AI constructs as well as from experimenting on the optimality, cost/effectiveness and quality of their designs.

4. PAYOFFS FROM AI SOLUTIONS IN THE TELECOMMUNICATIONS DOMAIN

Today, the pace-setting expert system for network design done by BBN in the mid-1980s has in many ways been superseded by more rigorous developments that incorporate both simulators and AI constructs. These mathematical models typically focus on a graphical environment and rapid prototyping of distributed networked systems and protocols.

Work on the integration of expert systems and simulators is done on the premise that designers of distributed networks must be given the ability to study systems operations in a variety of scenarios. For instance, they must:

- Analyze performance features and the dynamic response to the failure of switching nodes or links.
- Study the performance of a distributed message or transaction processing system under a variety of load models.
- Examine command execution and monitoring of communications aggregates of increasing complexity.

Such simulation and artificial intelligence systems must be user friendly and must employ graphical tools as well as monitoring media. Node functions and communication link behaviors must be easily coded by the user, with procedures linked to the simulated network model and executed efficiently.

The user must be in a position to reconfigure the simulation scenario, with the results of an execution graphically monitored. Although the original BBN model did not have the complexity years of experience suggest, its payoffs have been significant. Asked about the payoffs of DESIGNET and its offsprings at BBN, a senior executive underlined seven key points:

1. The speed of response to client requests.

There has been improvement of a magnitude order; it now takes 2 weeks instead of the 3 months that were typically required without the expert system construct. At the same time, productivity has increased many times over.

2. A better evaluation of alternatives.

This is a direct implementation of the McNamara principle: If we are going to do large investments, then we should have and examine many options—we should never box ourselves into just one possibility.

3. The conversion of different design approaches into achievable configurations.

The expert system tests at record time if the configuration is *doable*—and if it is, it prepares the bill of materials as well as the budgetary projections.

4. Customer satisfaction.

In a highly competitive market, customer satisfaction is of top importance. The customer must be given the assurance that the network under design will perform, and AI is instrumental in providing that service.

5. Credibility.

Every manufacturing company is faced with this problem and management well knows that if a company loses its credibility it is very difficult to get it back. "It is better to lose your eye than your name," says an ancient proverb.

6. The forecasting requirements.

An agile way of forecasting requirements permits a better focused network design with a longer life cycle. It also makes it feasible to cut down implementation delays, the biggest impedance in terms of network functionality.

7. A better network management.

Packet-switching reports on status, but such reporting requires an integrator, and integrative functions can be nicely supported through AI. We will look into this subject by briefly examining COMPAS and PROPHET by General Telephone and Electronics (GTE).

Contrary to BBN which concentrates on network project studies, GTE is an operating company interested both in the design of new telephone networks and in the steady maintenance of the established plants. GTE is also a leader in the implementation of AI. Four expert systems will interest us in this discussion: One of them is the Proactive Rehabilitation of Output Plant using Heuristic Expert Techniques (PROPHET). This AI construct:

- Addresses both cables and terminals,
- Assists maintenance personnel in trouble analysis,
- Helps control center staff in the identification of local loop problems, and
- Interfaces to GTE operating systems for preventive maintenance.

The analysis provided by PROPHET consists of five phases: Patterning, Testing, Analyzing, Dispatching, and Isolator Feedback. Among themselves, these phases automate the former manual- and DP-oriented identification process.

Known as 4-TEL, this elder DP-based process tests telephone lines overnight, getting a reading on every electrical fault (trouble) that it finds in the network. The results are collected by computer and stored in the so-called MARK system, which stands for Mechanized Assignment & Record Keeping.

By wisely choosing a hybrid approach that integrates DP and AI, GTE capitalized on work that already has been done. MARK is doing three things relevant to the PROPHET mission. It:

- Collects data from the past fifteen days,
- Converts telephone numbers to cable and twisted pair, and
- Outputs from groups of telephone numbers (called patterns) that pass a threshold.

For a number of years, a supervisor has been getting the list of 4-TEL and has analyzed this data to determine which patterns are likely to be bad cable sections. If a cable section is determined as being defective, a crew is dispatched to repair the bad cable.

By automating this process, the PROPHET expert system provides a faster response and cost containment. It both lowers maintenance costs and it offers GTE a competitive advantage through higher quality service. This project was also responsible for developing the COMPASS expert system which has been transferred to GTE Data Services for implementation.

COMPASS analyzes the diagnostic messages of a telephone switching network. It:

1. Examines maintenance logs from the central office, and it
2. Suggests maintenance action.

COMPASS is operational in many GTE telephone subsidiaries. It provides for a guaranteed maintenance of the existing switches by capturing expertise from the best telephone engineers of the firm.

Another GTE expert system CAF (Proactive Maintenance for Customer Access Facilities) identifies bad sections of cable that may need repair or replacement. There is, as well, work going on regarding machine learning for self-improving systems development, which will eventually impact all three GTE diagnostic expert systems we have considered.

Just as significant is a database-oriented development by the GTE Labs: The Intelligent Database Assistant (IDA). As an AI construct, IDA speeds and simplifies information searches by providing users with simultaneous access to multiple heterogenous databases.*

The company's Network Services and the Network Management functions rely on several databases that have evolved over time. Traditional database management systems provide the users with fixed-format query languages targeted exclusively to accessing a single database at a time running under one DBMS. This hardly answers the requirements posed by distributed database environments that prevail in a communications network.

IDA works along the principle of *federated databases* making it feasible to create a virtually homogeneous database out of many heterogenous ones. The outcome is an able solution to the perennial database access problem where heterogenous structures inhibit realtime retrieval of urgently needed information.

5. EXPERT SYSTEMS AT THE LAWRENCE LIVERMORE NATIONAL LABORATORY (LLNL)

The able management of distributed database facilities has become not only a major competitive advantage but also also a cornerstone to the effective implementation of information. Over the years, and this is true of all major organizations, databases have gotten fat, and management is throwing money at the problem instead of searching for solutions.

Part and parcel to this counterproductive policy is the buying of several vendors' DBMSs rather than the development of imaginative solutions like IDA, which are custom-made to the problem at hand. As a result, the company does not acquire the necessary skill, and with time, it falls further behind in its ability to deal with its sprawling data bases.

There are, of course, different viewpoints regarding this statement. Buying off-the-shelf software is more cost/effective than writing new routines. Quite often, however, not enough attention is paid to *homogeneity*, and the company ends with incompatible solutions it can ill afford.

Some organizations know how to use technology to their advantage. The fields in which avant-garde organizations operate are quite diverse and they include telephony, nuclear engineering, aerospace, and banking, but they do have in common the search for a better

---

* See also D.N. Chorafas, *Risk Management in Financial Institutions,* Butterworths, London, 1990.

solution than that provided by routine approaches, as the following examples help document.

The United States Nuclear Regulatory Commission (USNRC) has developed a knowledge engineering solution for the regulation of the nuclear industry: The Accident Management Expert System (AMES) is an operational aid to decision makers. Other AI constructs have been developed and implemented by the Lawrence Livermore National Laboratory (LLNL). Two groups compromise the artificial intelligence effort at LLNL:

1. The engineering research division focuses on a heuristic-based database management solution,* signal analysis and processing, and an impressive range of inferencing techniques.
2. The knowledge engineering group depends on the computer science department and it specializes in physical modeling, including the first principles, where appropriate. Its work is largely concerned with rule-based knowledge representation.

The overall approach of both efforts is characterized by combining artificial intelligence techniques with standard computational processing—and hence, with hybrid systems. The implementation philosophy focuses on integrating symbolic computation into standard processing methods.

Among the more notable AI projects at LLNL is a knowledge-based construct for representing distributed parameters (Figure 5-3), analyzing in qualitative and symbolic terms dynamic physical systems, developing planning models for open system implementation, and focusing on specific issues such as expert systems for triple quadrupole mass spectrometry, accelerator tuning, and so on. AI is also used in verification processes to:

- Monitor and interpret security alarms for sensed environments in the laboratory
- Control fusion energy experiments and the monitoring systems that they entail.

Magnetic fusion energy (MFE) experiments are conducted with large aggregates that make up several subsystems. The complexity of any experiment is such that most, if not all, of these subsystems must be operated automatically. To help in this effort, LLNL has developed an expert system that contributes to the automation of one of these subsystems: Neutral beam plasma heating, a major component of nearly all MFE experiments.

Before a newly manufactured or freshly overhauled power source will operate reliably at the experimental parameter values, it must be carefully conditioned. Conditioning is needed mainly because of the presence of contaminants such as dust particles on or near the high-voltage components inside the source:

- A dust particle will trigger arcing when the voltage is raised to a certain level.
- At high power levels, the arc will create a pit in the surface and will ruin the source.

---

* There is good reason to pay significant attention to intelligent databases. LLNL today has the largest database in the world with 9 terabytes operating online. It is shared by 11 Cray supercomputers.

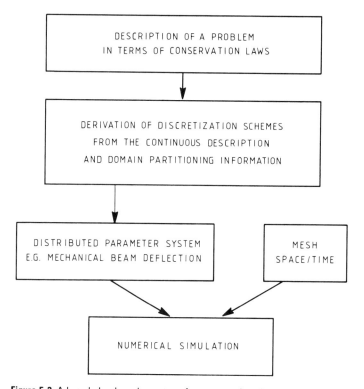

Figure 5-3. A knowledge-based construct for representing distributed parameters.

However, at lower power levels the arc will burn away the particle without affecting the smooth surface. Raising the voltage until arcing again occurs will burn away the next particle. The control process is called source conditioning.

Source conditioning may be viewed as a conventional supervisory approach in which the operator is the controller, but this is not automation. LLNL studies have documented that it is not possible to replace the operator with a conventional control system; replacement requires constructing a mathematical model of the source conditioning process and of the source. Doing so, however, is very difficult because the behavior of arcs and plasmas is highly nonlinear.

A plausible course is to emulate the reasoning and decisions of the expert human operator. The LLNL expert system sought to codify the operator's knowledge in a construct that operates (initially, at least) in cooperation with the operator.

The Accelerator Tuning Expert System (ATES) addresses itself to large accelerators and it makes use of a heuristic model to track various issues. The goal is to develop and refine the software methods through the use of knowledge-based techniques that involve the control, operation and tuning of large scale instruments.

The purpose of tuning an accelerator is to produce a beam that has the desired temporal and spatial energy and charge distributions for a given experiment. The most desired distribution, or beam profile, is presented in Figure 5-4.

On the left side, the beam pulse traveling down the beam pipe is within a few millimeters of the center. All electrons have uniform axial velocity and zero transverse velocity and are uniformly distributed within the pulse. In reality, however, the picture is the same as the right side of Figure 5-4, with the beam subjected to a variety of instabilities, so it diverges. The expert system's tuning goal is to produce a beam pulse propagating on-center down the beam pipe.

ATES focuses on the linear accelerator and its associated beam transport system, with two different approaches investigated similarly:

- A heuristic solution is pragmatic, tracking the work done by the expert technicians.
- A numerical, model-based approach is more theoretical. The lab's physicists focus on this one.

To deal with the continually changing status of the machine as component failures occur, the model uses AI concepts to reconfigure automatically the computer representation of the machine.

In order to map the large number of step-by-step procedures by which the operator tunes the various sections of the accelerator aggregate, LLNL knowledge engineers developed a representation called a Monitored Decision Script. Its key features are:

1. *Monitor rules,* which are periodically examined to detect changes in the machine,
2. *Scripts* that detail the steps needed to prepare the machine for an experiment, and
3. *Handler rules* that investigate the different exceptions that take place during the operation.

The LLNL knowledge engineers focused their efforts on two accelerators: the Advanced Test Accelerator (ATA) and the Experimental Test Accelerator (ETA), the former being the more complex of the two.

Another LLNL expert system addresses itself to triple quadrupole mass spectrometry. It has been developed over four years by a team of three knowledge engineers and a chemist. Its goal is to tune for optimal sensitivity, optimizing more than 20 parameters. This AI construct has achieved 90 to 150 percent of sensitivity by human experts in tuning, a much better focus than that possible through human operators.

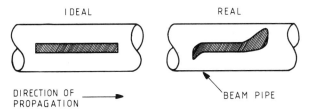

Figure 5-4. Beam profile while tuning.

This project is essentially a reflection of the fact that during the last twenty years chemical instrumentation has become very powerful and complex and it includes the acquisition and interpretation of data:

- In many cases, instruments require computer systems for their operation.
- In others, new instruments are too involved to be manually controlled.
- In some systems the data have to undergo processing by computers before they can be understood.

Instrument automation has traditionally been achieved using standard algorithmic methods implemented in procedural languages. However, as the LLNL researchers underlined, this approach is not suitable when the automation task involves poorly understood procedures and heuristics.

In a fuzzy environment, more powerful software tools are needed to explore different approaches in a timely manner. This has been, for instance, the goal of the expert system written to tune (optimize) a sophisticated chemical measuring instrument, the triple quadrupole mass spectrometer.

6. ALARM FILTERING, EMERGENCY CONTROL, AND OTHER POWER PRODUCTION EXPERT SYSTEMS

Designed by the Idaho National Engineering Laboratory and marketed by Bechtel Power Corporation, the Alarm Filtering System (AFS) is an AI construct that addresses the problem of the cognitive overload of power plant operators during plant upsets. Its goal is to bring the attention of the operator to those alarm signals that are directly related to the cause of the upset. This is accomplished by emphasizing those alarms related to the cause while de-emphasizing the cascading or nuisance alarms.

AFS utilizes object-oriented programming techniques to model the alarms of a plant and the causal relationships that exist between these alarms. It is currently being marketed as a retrofit to annunciator systems in control rooms for power plants. New microprocessor driven annunciator technology which requires a considerably smaller investment for installation and at the same time provides the basis for a more sophisticated solution.

A similar product is the Alarm Processing System (APS), a smart annunciator solution designed for the nuclear power plant control room at Diablo Canyon Nuclear Power Station. The architecture of this expert system:

- Reduces multiple alarm activity during transient operation, and
- Highlights unusual conditions which occuring during such transients which would not be detected using the conventional annunciator approaches.

Bechtel's TAI subsidiary has also conducted an evaluation of potential electric power industry utilization of Kennedy Space Center's systems autonomy artificial intelligence construct. The goal of this project is technology transfer because it has been part of NASA's program to investigate and demonstrate the applicability of space technology in the nuclear power industry.

The Kennedy Space Center's model-based autonomous reasoning model has been used to solve realtime sensor interpretations, diagnosis, reconfiguration, and control problems:

- In the nuclear industry, the initial application for this technology is control room alarm filtration.
- A microcomputer-based demonstration prototype has been developed, followed by a technology assessment and utilization plan.

Another of Bechtel's expert systems projects is SCOPE. It is used for the diagnosis and monitoring of power plant performance, based on the combination of rule-type reasoning, parallel with the steady simulation of the processes running in the power plant.

Several applications have been implemented with SCOPE focusing on the various water systems of power plants. One deals with the predictive maintenance of heat exchange systems. Another functions as flow balance assistance for the emergency feed water system. The first commercial application for SCOPE was the flow balance of the service water and emergency water systems for a domestic nuclear power plant.

Also by Bechtel, the Plant Performance Monitoring Systems (PPMS) is an AI construct that monitors a coal-fired power plant and recommends actions that an operator can take to optimize its efficiency. Working in realtime, this construct monitors parameters such as:

- Oxygen level,
- Carbon dioxide,
- Air flow,
- Steam flow,
- Opacity,
- Steam temperature,
- Condenser vacuum,
- Steam pressure, and so on.

After making various efficiency and controllable loss calculations, the expert system recommends corrective actions the operators can take that include air-fuel ratio control, boiler control, valve and slagging adjustments, combustion control, and so forth.

Designed for Electric Power Research Institute by TAI, the Reactor Emergency Action Level Monitor (REALM) provides assistance in identifying nuclear power plant emergency situations and the determination of their severity. It operates in realtime and embodies a hybrid architecture utilizing both rule-based reasoning and object-oriented programming. This application consists of over 300 rules, 1,000 objects, and 75 demons*; it took 6 man-years to develop.

The rule base consists of event-type rules and symptoms. The symptom-based rules go beyond the Emergency Action Level structure to address the more problematic scenarios and entail a symbolic representation of the plant information. In this context, REALM can:

---

* A demon is a supervisiory process. The term is used here in the same sense it is employed in Unix.

- Analyze plant conditions at the root level,
- Evaluate those conditions based on an overall perspective,
- Identify an emergency situation, and
- Classify its severity in relation to emergency action levels.

TAI also implemented an offline version of the REALM expert system to be used for computer-aided site emergency management training, known as Micro-REALM. Emergency planners also utilize this expert system to create experimental conditions and to respond to them. Instructors for emergency management training employ this expert system for sharpening the analyzing and decision making skills of in-training personnel.

The on-line version of REALM has been installed by TAI at the Indian Point 2 Nuclear Plant. Here the AI construct is used as a realtime operator advisor for performing situation assessment and suggesting the appropriate plant emergency status.

Another expert system, the Fan Vibration Advisor, assists an equipment repair technician in determining what is wrong with a malfunctioning fan. After the technician answer various questions about the technical details of the fan, the expert system calculates the vibration frequencies that must be compared to historical data by the technician. The cause of the malfunction is established by analyzing the vibration spectrum.

TAI also contracted with the Department of Energy to conduct an investigation, assess and demonstrate the feasibility of using artificial intelligence techniques for an Accident Diagnosis and Prognosis Aid to be implemented in nuclear power plants. A working scale model of the application was developed to provide an experimentation environment for rapid feedback on the concepts under investigation, including technical feasibility, practicality, and usefulness.

The successful completion of this project has demonstrated the ability of an expert system to diagnose an accident situation quickly and accurately. In its diagnostics, the AI construct provides insights into how an accident might progress, and it issues concrete, reliable recommendations on how to stabilize the plant.

6

Knowledge Engineering At Digital Equipment Corporation And At IBM

1. INTRODUCTION

That the old Electronic Data Processing (EDP) concepts can no longer serve the growing requirements of a knowledge society has been clear since the early 1970s. Less evident has been the solution which should be provided.

For more than a decade, computer manufacturers, and most particularly IBM and the so-called BUNCH (Burroughs, Univac, NCR, CDC, Honeywell), have tried to substitute mass for purpose. Not surprisingly, this ill-conceived effort failed, some of these companies going down the drain while they continued to show the EDP flag.

It is the users rather than the vendors who, in the 1980s, first found the way to the new landscape of AI-enriched computer applications. Among the vendors, the first to break out of the vicious cycle of applications obsolescence has been Digital Equipment Corp. (DEC). Initially, it did so to serve itself, and in rapid succession it also aimed to serve its customers.

As a result of the focused efforts in knowledge engineering which span a decade, DEC has developed a pallet of practical AI applications in a significant number of fields. These include network diagnostics equipment configuration, manufacturing operations, and also financial management.

One of the AI applications written to provide DEC customers with a competitive edge is the *Gatekeeper*. It has been developed to help Continental Airlines face gate assignment needs at Houston Airport. In operation since June 1988, the Gatekeeper recovered its costs in the first 2 months of operation.

Other applications have been made in the manufacturing domain. Scheduling is a popular issue and applications range from a variety of scheduler constructs for client organizations to a *Manufacturing Operations Consultant,* DEC's own implementation.

Other knowledge engineering constructs help the company manage complexity and sustain flexibility. *Email Browsing* is an AI application in office automation, screening and focusing on action regarding pending files. *Vax Performance Advisor* and *Enterprise Map* are management-oriented applications; the latter helps evaluate where expertise bottlenecks may lie.

Knowledge-Lens is an expert system integrated within more traditional computer software. This AI construct can be parametrically adjusted to operate at the system designer, team, enterprise, or community levels of business operations.

In this chapter, case studies are taken from active knowledge engineering development programs within DEC as well as from some constructs written for its customers. In a similar manner, emphasis will be placed on knowledge engineering applications by IBM who has recently tried hard to gain experience in the AI field in the USA, Europe and Japan.

2. DEC's THRUST INTO KNOWLEDGE ENGINEERING

DEC has given five reasons why it develops and uses expert systems: First, knowledge engineering is a strategic product; second, it assures the distribution of know-how and a better productivity; third, it makes feasible higher service quality; fourth, it enters into the growth market; and fifth it acts as a competitor in this field.

DEC claims that some 80 percent of all AI development work done anywhere is on its equipment. This, however, might have been true in the early 1980s and not today, though Vaxes are still among the more popular computers for AI applications, and DEC capitalizes on this fact for its market share.

The concept development by DEC is well exemplified by the so-called *Knowledge Network* which addresses itself to the handling not only of business orders but also of internal operations associated with them. The Knowledge Network follows the order cycle and Figure 6-1 identifies its major components as well as the way this system works. We will sample some of these AI components in more detail.

DEC says that the Knowledge Network places emphasis on the *business of the future* rather than on the factory of the future:

- Running on Vaxes, the component parts of the Knowledge Network help solve complex problems in order handling, machine configuration, scheduling, diagnostics, and maintenance, as well as in advisory, selection, and analysis problems.
- While these applications focus on DEC's internal operations domain, the system architecture itself is applicable to of the aerospace, electronics, motor, chemical, oil and gas, food and beverage, pharmaceutical, and telecommunication industries, among others.

One of the first AI constructs to be designed by DEC is the so-called *Decision Expert*. Its development dates back to 1981, when General Electric company initiated a program called GEN-X, on which the Vax Decision Expert is based. Since then, GE has developed dozens

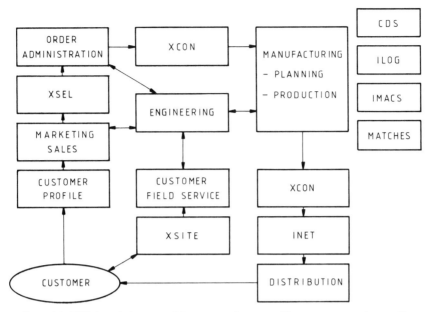

Figure 6-1. DEC's knowledge network interconnecting some of its expert systems in operation.

of successful applications on GEN-X, ranging from conventional diagnostics to statistical analyses.

The Vax Decision Expert can serve the needs of clerks, administrative personnel, technicians, engineers, and managers. The list practically includes anyone with:

- A diagnostic-type problem
- That calls for analysis, selection, maintenance, or help-desk advice.

In finance, Vax Decision Expert may be used to evaluate insurance risks and credit or loan applications. Financial experts at the State University of New York, at Albany, modeled after it an expert system called *Class;* it helps loan officers determine whether they should grant commercial loans.

Other AI implementations run on Vaxes are production-floor oriented. A manufacturing application at General Electric, for example, is *Machine Adjuster's Helper.* It assists engineers to maintain the complex machinery used in making light bulbs:

- An expert system focusing on the salient problem of a manufacturing industry significantly reduces costs.
- The proper maintenance of the machinery is crucial, and it has a direct effect on downtime experienced on the manufacturing line as well as on the number of faulty bulbs.

Another expert system program is *Care.* It helps in computer-aided requisition engineering for motors at GE. Given customer requirements for a motor, this AI construct can determine a number of factors in motor design to meet those requirements. It does so without having to call upon engineering experts, freeing up their time to deal with the specialized problems that only they can attack.

If GE is an example of a Vax user with significant AI applications, DEC itself, within its own operations, has put its equipment to work in an expanding range of knowledge engineering fields. The following list outlines some of the better known internal expert systems, though a couple of them will be further detailed in the subsequent sections:

1. *The Expert Configurer* (XCON) is an intensely used commercial expert system.

It was initially developed at Carnegie Mellon University as a 500-rule prototype, referred to as R1. Written in OPS5, this construct is employed on a daily basis in Digital's worldwide manufacturing operations to configure most Vax and PDP-11 computers that are on order and are not yet manufactured.

2. The *Expert Selling Tool* (XSEL) is an adjunct to XCON, helping the salespeople interactively configure computer systems to prepare correct offers and quotes.

XSEL's work can be done right in the customer's office. This expert system is also used to train salespeople, as well as to submit orders to XCON for further handling.

3. *Expert Site Preparation Tool* (XSITE) acts as a site-planning assistant.

It helps in the layout and design of a computer room. Typically, it runs after the components are selected, using XSEL as its reference.

4. *Intelligent Network* (INET) is a frame-based simulation system with knowledge about factories, warehouses, distribution centers, and the way to move material from one unit to another.
5. *Network Troubleshooting Consultant* (NTC) gives interpretative analysis and consultative advice on Ethernet/Decnet-related problems.
6. *Intelligent Scheduling Assistant* (ISA) schedules shop floor and manufacturing activities.
7. *Intelligent Business System* (IBUS) and its subset, *Intelligent Logistics Assistant* (ILOG), help with the execution of orders and order flow.

The expertise centers on problems in manufacturing. IBUS also teaches its users how to manage change and how to focus on people, rather than software, in a changing environment.

8. *Intelligent Diagnostic Tool* (IDT) assists technicians in diagnosing systems that do not pass manufacturing verification tests.

This expert system uses search techniques to locate failed modules within a computer.

9. *AI Spear* is a failure analysis tool for TU78 tape drives.

It detects intermittent failures by analyzing the system error log. Such failures are often impossible to find with conventional diagnostic programs.

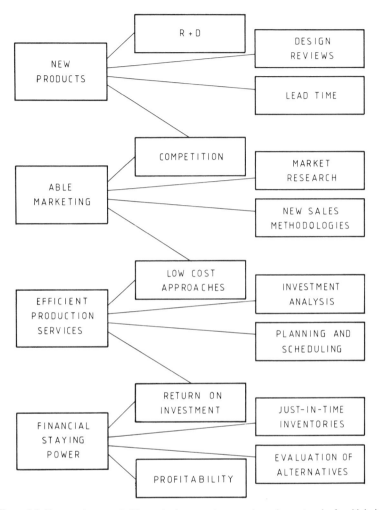

Figure 6-2. New products and able marketing must be seen through a network of multiple layers.

    10. *Knowledge-Based Test Assistant* (KBTA) is an expert system to aid hardware test engineers.

It implements a complete test programming environment from the acquisition of device information to the synthesis of test programs. DEC has also developed an expert system to enhance software functionality:

    11. *VMS Expert* (TVX) assists users of the VMS operating system.

It tutors users about utilities, commands, and functions provided by the OS; diagnoses error conditions the user may encounter; and helps to tailor sets of commands, converting the user's stated goals into VMS primitives.

What these references have in common is the astute utilization of recognition knowledge used to drive an expert system's behavior, provided that it is possible to determine locally (that is, at each step) whether taking some particular action is consistent with the acceptable performance of the reference task.

As the DEC and General Electric examples suggest, there is plenty of purpose for expert system usage by a manufacturing company. Figure 6-2 integrates different experiences in this connection to suggest fertile fields for implementation. In principle, when an expert system is implemented as a production system, the job of refining and extending its knowledge becomes quite easy.

The outlined range of knowledge engineering implementations is important for another reason. Today any major organization faces the fact that demand for experts has outstripped their availability. Hence, by relieving such a bottleneck, expert systems may be the answer to this challenge.

3. THE CONCEPT OF AN EXPERT CONFIGURER

Configuration is a process that helps determine the structure of a man-made system. A corollary problem is that of compiling without any error a complete list of parts as well as defining exactly how these parts should be interconnected, in order to assemble a working product.

In the general case, the parts list—or more precisely, the Bill of Materials (BOM), is a superset of the customer-specified parts. The customer, for instance, may have specified:

- Processor(s),
- Memory modules, and
- Communication lines.

However, both the salesperson and the customer may not have written down (and hence, failed to specify) the power supplies or the controller cards. The same is true of the interconnection of parts often expressed in terms of part placements requiring specification of the bus and bus-position for each card.

This example brings into perspective another crucial reference to systems configuration: *consistency checking*. In the old times, when man-made systems were simpler, this could be done by hand or simply assisted through classical DP. Today doing so is unwise because:

- The risk of error is too great,
- Configuration errors result in delays, and
- Delays cost dearly in money.

In essence, what every industry needs is an interactive expert system that takes a customer's marketing requirements and provides solution selling support for the salesperson. Its interactive features should as well be useful in clarifying the customer's own ideas about the outlined requirements.

In the case of DEC, this is done by XSEL to which reference has been made in section 2. As a sales-oriented configurator, it addresses both:

- A client's new orders for computer equipment, and
- A client's support orders that upgrade the existing machinery.

For instance, a customer who already owns a computer may want to add memory modules, communication adaptors or a vector processor.

As with the initial order, the expert system configurator must check for system structure and part consistency, assure that prerequisites are satisfied, and then do part placement to determine the entire BOM as well as to construct an assembly description.

This is more complex than configuring a new system, as it presupposes that the AI configurator knows what is already installed as well as which existing parts must be moved in order to make new parts fit. It is technically desirable to minimize the movement of existing parts, especially ones that are difficult to move around or may result in failures.

Often, the answer to these questions is complex as customers may purchase and install parts from different competing vendors. In other words, it is not simply a matter of accessing the client data base maintained by the computer manufacturer in order to know the existing BOM by customer and/or by system. External information may be needed as well.

In its original version, XSEL addressed itself to new configurations, with its 2,000 rules helping to specify a complete order, including a room layout. As stated, XSEL is connected to XCON providing a consistent and accurate BOM description.

"Ten or 15 years ago we were hiring for salesmen young engineering graduates, IEEE members," said a senior DEC marketing executive. "They were wizards in configuring systems, but not so brave in selling them in a highly competitive environment. We don't hire these people anymore; we hire people who know how to sell in a very competitive sense—but they don't know how to configure a computer. This is done by expert systems."

DEC addressed this issue by getting its best computer configuration specialists to express their expertise in terms of rules. These rules became the expert system's knowledgebank. Then it trained its salespeople to use in using XSEL which is the *productizing and fielding* phase of expert systems development.

How well has this effort gone? Today, an estimated 50 percent of DEC's sales engineers have been trained in XSEL—but only half of them (or 25 percent of the total) use it. The lesson is simple: Never underestimate the cost of producing an AI construct and selling its usage to the population for which it is intended. Implementation is never easy!

Taking a different but similar example as a reference, Honeywell's *Mentor* expert system was also developed in the mid-1980s and written as a configurator:

- Its development cost has been $100,000 and it took less than 1 year's work to develop.
- But its production cost was $1 million, and it took one full year to bring to the field.

Productizing includes documenting, training, finding ways to bring the AI construct to the field effectively and selling the wisdom of its implementation to the sales force. In short, it involves *getting the AI system accepted.*

The benefits a manufacturing company can derive from the use of AI technology are one thing; how to convince its people that such benefits warrant the expert system's usage can be another, and this is true even when such benefits are real. XCON, for example, reduces the DEC's sales representative time in configuring a client offer from 3 hours to about 15 minutes, a change that is more than one order of magnitude. But how do you sell the use of the AI construct to 100 percent of the sales force?

4. XCON's DOCUMENTED BENEFITS

The XCON challenge was created as a result of a marketing strategy: Many one-of-a-kind computer systems (Vax, PDP-11) were ordered and something had to be done to automate the handling of information relating to this equipment:

- Computer order processing requires a large time- and resource-intensive configuration task that involves technical edits of orders.
- By the late 1970s traditional methods had reached their limits and could not be relied on to help cope in an able manner with the growing number of customer orders.

Today, XCON is the largest and most intensively used commercial expert system, starting as DEC-sponsored Project R1 at Carnegie Mellon University where Professor McDermott developed the initial 500-rule prototype.

Originally written in Bliss (a language based on OPS5), XCON was taken over, restructured and enlarged by DEC. It is presently used on a daily basis in Digital Equipment's worldwide manufacturing operations to configure most Vax and PDP-11 computers that are on order and not yet manufactured.

XCON works with a large database of computer components. Its rules determine what makes a complete order.

- They check the bill of materials,
- They define spatial relationships among components in an order, and
- They help establish a process of configuration design.

XCON is an example of *a synthesis-oriented* expert system, in contrast to Stanford University's Mycin which demonstrates analysis-based approaches. XCON's job is to select and arrange the components of a computer system:

- The CPU,
- The memory,
- The terminals,
- The tape and disk drives, and
- Other peripherals.

Configuration is an excellent application for an AI construct because it requires the tracking of hundreds of options that can then be matched in almost limitless combinations. XCON moves forward, executing one rule after the other until it has accurately configured a computer system out of as many as 200 or more components.

The development history of R1/XCON helps to illustrate the problems encountered when an expert system is used in a practical setting. Since the Vax and PDP-11 computer lines were constantly upgraded and expanded, it was continually necessary to add more knowledge and greater capability to the knowledgebank:

- As a result of this activity, before long the R1 expert system became complex enough so that it was necessary to reimplement it entirely.
- For support purposes, DEC established a special software support group that had to invest considerable effort in understanding CMU's development before they could make any meaningful changes to it.

Today, a new copy (a minor release) of XCON reaches manufacturing and dispatching operations practically every 3 months. This calls for an elaborate update and training procedure.

At the positive side of the balance sheet, however, the application has brought significant benefits to DEC—and this has occurred since the beginning of the implementation. After the first three years of XCON usage, as of 1986, DEC realized, according to its own statement, the following profits from XCON:

Year	Money
1983	$18 million
1984	$25 million
1985	$40 million

At that time, i.e., 1986, including its associated subsystems, XCON had about 5,000 rules and represented an investment of $17 million. These 5,000 rules managed more than 12,000 Vax/PDP components.

Many people ask why this configurator job could not be done in a more classical way. To the query, "Could these programs be written in Cobol?" DEC answered during our meeting: "Yes, but:

- They could only do part of the current job, or roughly 1/3 of it,
- This would have involved an estimated 750,000 Cobol statements, and
- Such an effort represented 125 man-years."

In other words, if classical programming approaches were used, less than 1/3 of the configuration job would have been done and the resulting programs would have cost three times as much or more. The difference is of an order of magnitude in favor of the expert systems solution.

A Classical DB approach can meet neither the *sophisticated requirements* nor the *massive dimensions* of present-day computer-based jobs. One of the leading Japanese banks was to state, during an April 1991 meeting in Tokyo, that if classical DP solutions were used the accumulated backlog would have amounted to an astonishing *70,000* man-years of work.

DEC also claims that without XCON today it would have needed 2,000 people to do the configuration job. Through classical DP methods, some of the associated applications from advanced billing to cost control, would not have been made with the accuracy with which they are now done. What is more, XCON has had significant synergy effects.

In engineering configuration solutions, XCON has *increased the productivity* of doing design work by a factor of 100. "Even if it was a factor of 10, it would have been a revolution in engineering design," said an executive.

- The average time to configure a Vax order is 2.3 minutes with XCON, and with PDP-11 it takes less than one minute.
- For expect technical operators, it used to require 30 minutes to 30 hours.

Most importantly, human experts got about 70 percent of the configurations correct. The hit score with XCON is 98 percent.

Critics are saying that XCON has become a maintenance nightmare, and it is poorly understood, badly structured and, hence, hard to change. Some of these critics must have been born last night and have never heard what software is. We have not reached even a reasonable level of dependability with classical software and, for the time being at least, AI has not addressed this subject—though it should.

Software is the soft underbelly of computer science, and this is just as valid of highly sophisticated programs as of simpler subroutines. All things counted, AI programs are easier to maintain (more precisely, to *sustain*) than classical software; this, however, does not mean that the sustenance perspectives have been addressed in an able manner as should have been the case.

The need for sustenance is real. To help maintain the hit rate just mentioned in Vax configurations, the knowledgebank of XCON is in a steady update. Table 6-1 presents comparative growth. An estimated 30 percent of the rules in the knowledgebank change every year.

This can be said in conclusion regarding benefits obtained from XCON: There has been a very significant reduction of configuration time and an even more impressive reduction of configuration errors. There have also been better logistics at the assembly line, highly improved system tests, faster results in terms of configuration debugging, an accurate maintenance of configurations and, as well, a redistribution of materials and components.

5. A DEVELOPMENT TIMETABLE AND ITS CHALLENGES

Configuring a computer system to customer specification is in itself a demanding job. Therefore, the project that started in December 1978 at CMU was demanding. Yet, within 4 months, in April 1979, a prototype with 200 rules was demonstrated. Less than a year later, in January 1980, the use of the expert system with real life orders began.

TABLE 6-1 Statistics on XCON Growth

	1980	1986	1990
Component database Parts description	400	12,000	40,000
Attributes by component	8-10	15-20	25-125
Computer families	1	10	30
Number of rules	700	5,000	15,000

DEC assumed the development of XCON in January 1981. At about that time, the XSEL development started with a prototype demonstrated in July 1981. One can appreciate the fact that even in those early days of practical AI efforts, the pace of development was fairly fast.

However, as the preceding sections have underlined, it is easy enough to build the initial prototype and not that difficult to develop the expert system. More tough is to train people about its usage; in this, XCON has been much more successful than XSEL has been. One of the reasons may lie in the fact that the procedure adopted closely emulates what was done prior to the AI implementation, but it automates that activity:

- XCON accepts as input the list of items on a customer order.
- It configures them into a system.
- It notes any additions, deletions, or changes needed in the order to make the computer on order complete and functional.
- It prepares a blueprint of detailed diagrams showing the spatial relationships among the components as they should be assembled in the factory.

Before XCON and prior to the complications presented with Vax clusters, technical editors did the handling of orders. This has now been thoroughly automated as XCON provides all the necessary pieces of information per order while it creates a usable configuration. Such an automated process performs about six times as many functions as the technical editors used to handle.

XCON has allowed Digital Equipment to avoid costs that it would have incurred if it had been forced to hire more technical editors as the volume of computer systems sold increased. Before XCON's accurate configuration plans were available, computers were sometimes assembled up to the point at which a problem was discovered. When this happened, the Vax or PDP 11 had to be dismantled and reconfigured.

While these problems have been solved with XCON, others have come up, for instance, the difficulty of testing an automatic configurator. This primarily relates to the domain rather than to the use of expert systems tools:

- Evaluation of an output configuration requires trained personnel to assure that relative and absolute constraints are met.
- This is particularly tough for the set of constraints since their validation calls for a most detailed search that is manually performed.
- Another difficulty is encountered in regression testing. This becomes necessary as changes happen to the Configurator.

Remember that the *sustenance* of an expert system requires a steady upkeep of the rules. Since the configurator performs an exhaustive search, changing one rule or constraint may cause the exclusion of many configurations that are wanted or, alternatively, the inclusion of some that are unwanted. As a result, part placements may be different from those computed before *sustenance:**

---

* As already underlined, when we deal with the maintenance of expert systems we talk of sustenance. A knowledge engineering construct is sustained, for instance by uprading its rules, rather than maintained.

- Small changes in the relative and absolute constraints can cause large changes in the components placement, while
- New constraints on part placement may require further additions to the Bill of Materials list.

True enough, configuring a computer product manually is time consuming and error prone. We *must* automate the configuration process, but we must not forget the work this process involves.

In terms of the assembly of complex computer systems, a configuration is acceptable if it complies with two kinds of constraints:

- Absolute constraints, the conditions that must be met in order for the projected configuration to be valid, and
- Relative constraints which state that some characteristic of a configuration should be optimized.

Since absolute constraints reflect the configured computer's feasibility, they should never conflict with each other. In contrast, relative constraints may frequently appear to be in conflict. In reality, what this conflict means is a universally acceptable prioritization of relative constraints, permitting a heuristic implementation.

Typically, such implementation will be done according to the rules in the expert system's knowledgebank in a way apparent to its users. In XCON's case, the endusers typically are:

- Technical editors who are responsible for seeing that only configured orders are committed to the manufacturing floor.
- The technicians in Digital's manufacturing organization who assemble the computer on the plant floor.
- Salespeople, who use XCON in conjunction with XSEL, as the latter helps them prepare accurate quotes for customers. This can be done on a dial-up basis from the customer's site.

Other users at the plant site are scheduling personnel who employ information from XCON to decide how to combine options for the most efficient configurations. All these users benefit from AI technology.

Finally, we should be sensitive to the need of integrative perspectives from the sales order to the assembly description, which is the detailed plan used by manufacturing to construct the computer. Such an approach with the appropriate methodology is necessary to accommodate a variety of parts and components handled in a modular way. This means an ambitious development of the knowledgebank made to:

- Separate the control strategy, encoded as *metarules*.
- From the specific domain knowledge which is expressed in rules.

The rules that contribute to a particular goal in a knowledgebank have to be grouped. This makes it feasible to have the AI construct partitioned into several subtasks, as the XCON reference helps document.

Lessons from the XCON experience have to be learned now and serious thought must be given to the necessary evolution of expert systems applications. For instance, since complex expert systems generally use a large number of computer resources, the knowledgebank and the working memory can be separated so that each can be independently processed on parallel processors. Such a separation can also make the expert system easier to change.

6. IBM's THRUST IN AI

IBM, too, has an expert system configurer that we will review in this section, but the mainframer's first in-house development to hit the public eye was the Yorktown Expert System for MVS, better known as YES/MVS. A similar project done by IBM Japan in Kawasaki received scant attention, yet it has been much more able to run 12 fully configured large scale systems by residing in one of them.

What both MVS and the Kawasaki solution have in common is the automation of computer operator chores. To perform their job, operators maintain a conceptual model of the computing facility which is:

- In part static, as the current hardware and software configuration are, and
- In part dynamic, involving the job stream, the state of key system indicators and the status of specific or potential problems.

Like human operators, YES/MVS maintains a model of the target computer, including the status of devices and the current state of particular problems as well as the time stamp of information it has available. Since YES/MVS interacts with human operators, its rules reflect on the operators themselves, including what requests they have received (from the expert system) and what responses they have made.

In other words, YES/MVS operates by soliciting status information from the target system through commands. It interprets data, queries the operators, emulates the resulting reply or lack of it, and handles multiple responses to a single question.

In the first version of YES/MVS, each subdomain independently maintained its own model of the target system to minimize the need for coordinating the activities. However, this led to a duplication of effort and to the realization that even if status information is not shared among subdomains it is still worthwhile to have a common service for housekeeping.

In essence, the first version of YES/MVS was a research prototype that provided facilities for:

- Managing the queue space,
- Scheduling large batch jobs,
- Handling problems in channel-to-channel links, and * Assisting in resolving hardware.

What IBM has discovered from this experiment is the importance of expressing a knowledge of computer operations in a general manner if expert systems are to be easily moved from one installation to another for automating computer operations. For channel detector errors, for instance, YES/MVS accomplished this feat by having two types of rules:

1. Expressing strategies for isolating errors based on installation-specific characteristics, and
2. Dictating actions to be taken when a given type of component fails.

Another lesson learned from YES/MVS is that expert system shells for automated operations should provide an integration of rules and procedures, as well as support for interprocess communications, timed reminders and customized conflict resolution. Language constructs, as well, are necessary for easy adaptation.

Another lesson learned regards the wisdom of extensively testing expert systems for automated operations. Such tests must be polyvalent and done in a statistically valid manner, under different operating environments.

As stated in the introduction to this section, IBM has also built an expert configurer. This expert system is today used worldwide for virtually every 9370 computer. Covered operations include:

- Customer proposals,
- Orders, and
- Order upgrades.

IBM is constantly enhancing this 9370 configuration expert system as the product offering changes. Through it and YES/MVS, the company learned much (both bad and good!) about the development of a large complex expert system and how to integrate it into the business flow of established, traditional product lines.

This aftermath is an important, often overlooked aspect of the expert systems experience. The 9370 configurator today handles in excess of 10,000 sessions per month initiated by IBM marketing personnel around the world.

Encoding relative and absolute constraints in rules facilitates the sustenance process though changes are not always easy and testing remains difficult as upgrades have to be steady. One of the lessons learned is that a configurator:

- Should recognize its own limitations, and
- Should request operator intervention.

This is an issue serious users of AI constructs should keep under perspective when major projects are contemplated, particularly when it is known that they should have a fast changing knowledgebank.

YES/MVS, the 9370 Configurer, and CASES (which we will consider in the following section) are targetted at computer applications which will be of strong interest in the future. They all have been served through a small group of top level research workers, and they focus on a few technology areas in which specialized expertise can be developed from operating experience.

At IBM, as with other computer manufacturers, applied AI research projects are chosen to address specific, timely issues with a vision of the directions of computing over this decade. Some of these projects operate as technical information brokers, providing channels among universities and between IBM's research department and universities.

They also aim to provide a systematic approach to the difficult problem of reconciling:

- The information technology infrastructure of an organization
- With its strategic objectives.

The goal is to help the user get the sense of the information he is confronted with, on the premise that most material is not written to provide the reader with essence and understanding for decision-making purposes.

An example of a successful effort in this direction is the Capital Asset Expert System (CASES) which, in the mid-1980s, became the first application of AI to go into production mode in an IBM facility. CASES is also a significant application in terms of the results that have been obtained. The key issue here has been one of taking high technology and applying it, in a pragmatic down-to-earth manner to real-life business problems.

7. THE CAPITAL ASSET EXPERT SYSTEM (CASES)

Developed by IBM industrial engineers at the Endicott factory, CASES has aimed to simplify various operating procedures in business management that include ordering, acquisition, transfer to/from, storage, utilization and disposal of assets. In so doing, the expert system gives all-encompassing advice.

Let's start with the background of this implementation. In IBM factory jargon, *Capital Assets* is a problem area that involves a considerable amount of business operating procedures, precisely those which:

- Need to be followed carefully to make the company run
- But are quite often complex or even contradictory.

Used in an "assistant to" mode during the daily routines at the factory, this expert system advises IBM employees on how to deal with established business procedures:

- Cutting through different types of bylaws and directives,
- Providing comprehensive answers over the whole interdepartmental area of procedural application,
- Integrating signature authorizations for transaction validation, and
- Avoiding wasting time—but also reducing the industrial engineer's level of frustration.

Today, the Endicott factory employs some 800 industrial and manufacturing engineers who spend some of their time doing what the expert system can do. An *asset,* for instance, a machine tool, is received, installed, expedited, shifted and disposed of. In each case, considerable paperwork is involved, and while Endicott has been rebuilt and fully modernized several times, some of the procedures date back to 1926 when it was first built.

The users of CASES are the 800 industrial and manufacturing engineers, and its implementation has had significant results in terms of management overhead. The Endicott engineers' and managers' appreciation for the expert system is seen by the fact that the end users are its real promoters once they get the prototype and test it.

Several projects were considered at the Endicott factory prior to this implementation, but the Capital Asset area was chosen for a number of reasons. One of them is that existing regulations and bylaws confused people. As a result, many forms were filled out incorrectly, lacking the correct signatures, and so on:

- The department controlling these forms was questioned constantly over the phone and in person with the "how to do" kind of queries.
- At the same time, the answer to these queries was given, but it was not codified. It was also not possible to exploit in a batch manner or through Coboloid-type programs.

In other words, the factory possessed thorough, perhaps confusing, operating procedures defined for asset management requirements. They simply were not easy to identify or to follow, but a few people in the asset control department knew what those operating procedures really meant and how to interpret them. There were problems, however, in getting needed information when these people were not available.

Figure 6-3 shows the procedural graph applicable to Endicott operations. Assets get moved, transferred, physically scrapped. But at times the paperwork can go poorly and people give up trying to find information.

In contrast, with an interactive expert system most of the consultations are solvable in just a few minutes. This was evident to the designers as was the fact that the project seemed doable and not overwhelming:

- Asset management lent itself to easy decomposition in function.
- Hence, one can fully develop a small problem area for an initial prototype.
- But the designer can also easily add on his or her expertise at a later time, doing so in a modular manner.

CASES guides the hand of the IBM manufacturing engineer not only in the selection of forms to be filled out for asset management but also in the identification of appropriate boxes within a form. This can be a confusing and time-consuming job when done manually through guesswork. For instance:

- One of the forms had 47 boxes on it, but only *one* was needed for a given type of asset.
- Some forms required up to seventeen signatures per box, while sometimes only 2 were needed for another type of asset.

The solutions provided by CASES are available through online consultation, a dialog that assures in real time the provision of information pertinent to the user. Once the whole asset management work was converted to interactive video, it became possible to:

- Merge procedures and forms from many sources,
- Create a factory knowledgebank for written operating procedures and directives,
- Guarantee a unique ability based on symbolic entities,
- Assure one cohesive course of action, customized to each individual situation, and
- Obtain system availability 24 hours per day, 7 days a week.

In this particular application, expert system components include 33 parameters, 653 rules, 34 focus control blocks, and 34 custom screens. There are also 3 external interfaces and 3 command files.

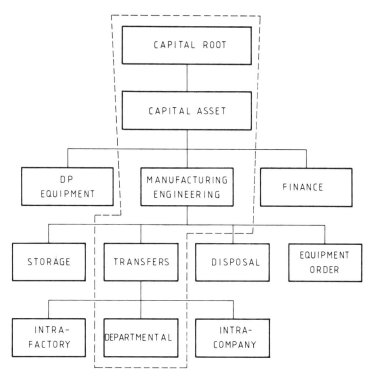

Figure 6-3. Capital asset allocation is a multidimensional proposition.

CASES runs on the 3090 and 3081. Its development was fairly fast requiring 6 man-months of effort spread over about 1 calendar year because of delays in module evaluation. The cost amounted to a very affordable $42,000 while the estimated payback to IBM has been at the level of $1 million within its first year of implementation.

Throughout the project, the development process was most methodological, focusing and refocusing on forms, paying attention to the symbolic rules, and accounting for domain understanding. Also, user acceptance was never underestimated.

Final acceptance of the domain experts was gained through demonstration of results involving fairly complex scenarios—the hits being calculated by a statistical sample of experts, every time the expert system responded correctly to their requests. The testing phase also established that CASES compared best with the best expert's own results.

Pre- and *post*-consultation functions assisted the user in running a CASES session. The user can save the last consultation under a name picked by the user, save the screen images presented in the last session, view the CASES User Guide online, as well as contribute comments and suggestions online to the CASES developers.

A smart startup utility was placed on a common disk that all users at the site access automatically at log-on time. Its commands are essentially all the user has to know to access the application. Most significantly:

- The response time for the average query is a subsecond.
- The longest query on record has taken 3 seconds.

For sustenance purposes, domain experts got trained in knowledge engineering and software development to understand the importance and complexity of the maintenance function. This is consistent with the fact that a good problem for an expert system application is one in which there is agreement on the facts of the problem domain no matter how fast such facts are changing.

Future enhancements include modules for financial expertise, electronic signatures, data base tracking and the monitoring of equipment. Another advantage includes a tie into asset logistics and control, as well as the inclusion of other asset types than those presently supported.

Experience with CASES also provides documentation on expert systems' limited portability between different sites. Impressed by the obtained results, another IBM factory asked to use the AI construct within its own operating environment only to find out that this could not be done; it was necessary to reproject the expert system within this factory's domain and rules, though the methodology developed by the Endicott experience was transferable.

7

The Impact Of Second Generation Solutions

1. INTRODUCTION

When we talk of undertaking expert systems projects, we should be aware of the fact that project goal, size and format differ widely. They run from simple to complex; from $10 thousand to $1 million or to several million dollars; from using one or two knowledge engineers to having a dozen or more; from PC to supercomputer implementation. Also they run from rule-based expert systems of the IF...THEN...ELSE type to sophisticated second generation solutions.

What all AI projects have in common is that the emphasis is on the *solution*, not on some vague theory. At the same time, if the solution is not implemented, problem solving is just daydreaming. Hence, timely implementation is another positive characteristic of AI projects, though the methodology used may differ widely.

Speaking then of practical down-to-earth implementation domains with the overriding policy to get *results*, not simply models, it makes sense to talk of some developing trends:

> 1. The broader domain of artificial intelligence-enriched solutions, not just expert system models, will dominate the future of computers and communications perspectives.

In software and hardware architectures there is a starting revolution that can be expected to complete its current cycle by the mid-1990s. Its basic characteristic is that of hybrid systems integrating existing investments in DP with new AI developments.

> 2. Within the 1994 to 1997 time frame, data processing as we know it today will disappear and AI will lose its distinct identity.

As the decade of the 1990s progresses, value differentiation will dominate in the manufacturing industry and will be based on a new generation of on-line systems. These will utilize networked supercomputers, distributed databases and powerful workstations, enriched with artificial intelligence at most levels of business operations.

3. On the communications side, the integration of major achievements into *one logical network* will reach clients and suppliers at their workplaces.

By the mid-1990s corporate strategy will revolve around the concept of *solution selling*, leading to the merger of formerly distinct disciplines. This means spousing high technology as such with all its ramifications, not just rule-based expert systems.

Let's always recall that—as Chapter 3 underlined—expert systems are just one of the first practical implementations of artificial intelligence, and as such they are the present solution. More will follow. Nevertheless, while we should plan for the future, understanding the present is very important because the gateway to the future requires changing our culture. To do so, we need a transition period and knowledge engineering can help in this direction.

2. THE CHALLENGE OF FUZZY ENGINEERING*

In its fundamentals, fuzzy logic is a mathematical process that can draw a conclusion by judging fuzzy, ambiguous information. It does so in a manner similar to that of the inference process of human beings.

Whether through software or hardware (fuzzy chips), fuzzy engineering, in reaching a conclusion, helps judge the meaning of indistinct information, such as "rather big" and "almost zero." Applications range wide from the automotive industry, to banking and space research.

NASA has incorporated fuzzy chip technology into the space shuttle to help it control its altitude in space and to conduct docking maneuvers with another spaceship. A fairly similar implementation of fuzzy chips also opens the way for the development of:

- Industrial robots capable of smooth movements similar to those of human beings, and
- Sophisticated speed controllers for use in transportation vehicles in which passengers do not feel any vibration at all.

Fuzzy engineering solutions are designed to speed up information processing time greatly, substituting hardware for software. Taking the architecture of one of the developed fuzzy chips as an example, it consists of two elements each measuring 3 cm by 3 cm—the rule chip and defuser chip.

A rule chip designed at Kumamoto University, in Japan, is capable of conducting one million fuzzy calculations per second leading to a correct conclusion based on fuzzy set

---

* This theory was first presented in 1965 by Dr. L.A. Zadeh and Dr. R. Bellman, then of Stanford University.

theory. The defuser chip converts the conclusion drawn by the rule chip into analog numbers.

Fuzzy inference programs are already incorporated in conventional computers to provide a smooth speed control for subway cars, for instance in the city of Sendai, in northern Japan. But since, as any software, this tends to slow down the machine's original calculating speed, the fuzzy chip is a good solution. This statement is further enhanced by the fact that engineering uses logical process principles that are different from those used in classical computing.

The mathematical principles of fuzzy engineering reflect more closely real world events than classical mathematics do. They do so by incorporating and manipulating *vagueness* and *uncertainty* which often dominate many business propositions.

Industrial products, for instance, have life cycles largely based on uncertainty reflected through response, more competitive breakthroughs from other R+D laboratories, and simple wear out of the lustre a new product had. Based on research I conducted in the mid-1960s on changes in product vitality, Figure 7-1 exemplifies this reference.

The manufacturing industry tries to capitalize on the properties of fuzzy engineering and its ability to represent the uncertainty embedded in business systems. An implementation based on a joined project by Hitachi and Nissan Motors incorporates:

1. Fuzzy inquiry, and
2. Quick inquiry functions.

As a result of this dual approach, an inexperienced person can readily retrieve part numbers from automotive inventories. *Fuzzy inquiry* means that the exact part number is retrievable simply by entering a few items of car model information such as:

- The classification code employed on the automobile inspection certificate,
- The model/type marked on the model number plate,
- The general designation, such as body style and so on.

Quick inquiry focuses on nearly one million information elements that can be rapidly retrieved in a priority manner.

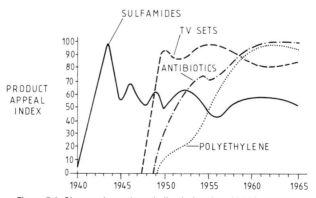

Figure 7-1. Changes in product vitality during the mid-20th century.

Another impressive application of fuzzy logic is in the diagnosis of engine troubles. In a fuzzy engineering project done at the Tokyo Institute of Technology, Professor Toshiro Terano and his assistants investigated the reports of engine trouble of 120 large merchant ships over three years. This project focused on 89 items such as the location of failure, kind, damage, detection method, estimation, diagnosis, counterplan and economic loss. As a result of this analysis, it became possible to identify:

- What kinds of trouble can most effectively be avoided, and
- What kinds of performances are expected for the safety system.

This phase has been followed by the development of more rigorous heuristics for diagnosis through fuzzy logic. This approach is based on the premise that human perception and decision are very subjective and fuzzy, yet they play the leading part in the case of trouble identification.

In this safety system, the signals from the mechanical sensors and from the human feedback are processed together. For this purpose, a new inverse operation of fuzzy relation has been developed. The results of diagnosis are presented online through soft copy as short sentences that tell the operator clearly or vaguely the:

- Locations and causes, as well as
- Countermeasures for engine troubles.

One of the important results obtained from this study is the classification of engine troubles. All kinds of troublesome factors can be classified into five groups, G1-G5, as shown in Figure 7-2. From the physical meanings of components and the allocation of factors, the researchers deductively stated the characteristics of each group in the following terms:

Group 1 (G1): Electrical and mechanical equipment characterized by the *perceptive* detection of incorrect performance or locking.

Group 2 (G2): Rotating machinery, including the troubles of pumps and air compressors, characterized by the detection of abnormal noises.

Group 3 (G3): Boilers and automatic control devices whose failure is detected by observing meters such as pressure and temperature.

Group 4 (G4): Other electrical machinery whose troubles are detected by the human sense of smell.

Group 5 (G5): Main engines, represented by diesel motors and located near Group 3 in space, whose damage is very large if it happens, but diagnosis is a little easier than in boilers, so the percentage of correct diagnosis is rather high.

The fuzzy relational equations developed at the Tokyo Institute of Technology by Professor T. Terano and his coworkers provide a model for describing the causal relationship between failures and symptoms. Possibility theory has been used reflecting the fact that:

- A failure may cause all the symptoms relating to it.
- A symptom appears when any one of the failures relating to it occurs.
- When two or more kinds of failures occur simultaneously, the accuracy of a

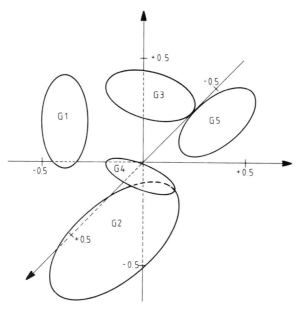

Figure 7-2. Dr. Terano's research on clustering of engine troubles through fuzzy engineering.

symptom is determined by the maximal value of those brought about by each failure.

Based on this work, the researchers developed a new safety system from the viewpoint of systems engineering. This solution works as an intellectual assistant to the human operator. It detects and diagnoses engine troubles with the help, at an early stage, of the perception of the operator. By using fuzzy logic, the results are presented in a suggestive way to the worker—which is the way in which he operates.

This is admittedly a difficult passage, the challenge being *conceptual* and *cultural* rather than technical in nature. But there is by now plenty of evidence that the Zadeh, Bellman and Terano fuzzy theory can be effectively applied with significant success to mechanical control, process engineering and basic man-machine interactions—this being done in an easy-to-implement, user-friendly manner.

1. Pattern recognition
2. Object identification
3. Robot vision
4. Character recognition
5. Signature recognition
6. Flaw detection
7. Fraud identification
8. Production scheduling
9. Quality assurance
10. Constraint optimization

11. Prediction of node failures, and
12. Emulated behavior

In manufacturing, neural network projects include artificial vision, quality inspection, precision alignment, as well as VLSI design, intelligent management of the production line, and the development and use of autonomous vehicles. Neural constructs are successfully used in telecommunications in applications fields, which include hardware configuration, call scheduling, call routing, and troubleshooting.

Among neural network applications in mathematics we distinguish, number theory, function analysis, the solving of differential equations, statistical analyses, and optimal grid transformations. Also included are a growing number of optimization problems such as the travelling salesperson. The development space of a neural network is shown in Figure 7-3 and it runs along three axes of references:

- Type of model,
- Hardware, and
- Software and firmware.

In a recent neural network project regarding the recognition of handwriting, the inputs have been data vectors representing character strokes from different people. The output is a fairly crisp "A," "B," "C" . . . "1," "2," "3" . . . Single-digit character recognition is now being achieved with an accuracy of 90 percent to 95 percent. More difficult is the recognition of continuous sentences where the accuracy stands at a low two-digit number.

To evaluate the behavior of the neural network circuit in a character recognition application, the experimenters developed an interface connecting the chip to a computer. Data

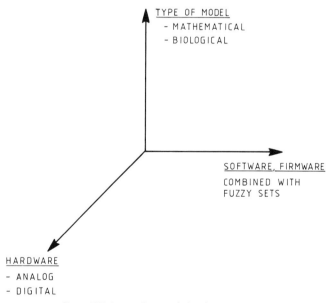

Figure 7-3. A neural network development space.

transfers were made directly from the computer's memory to the chip to handle the flow of the large amounts of data characteristics of image processing requests.

One complete processing cycle that included loading the input vector, accomplishing a computation with the circuit, and reading back the result into the computer, required approximately 25 clock cycles, but most of the time was required for reading the data in and out. The whole process of recognizing a character proceeded according to the following steps:

1. A handwritten character is read with a camera, digitized, and then normalized in size to fill a 128 x 128 pixel frame.
2. The image is coarse blocked into a 16 x 16 pixel binary frame.
3. The skeleton of a character is structured with the width of the lines reduced to one pixel.
4. This skeletonized picture is searched for a number of geometrical features.
5. The positions of such features are compared to a training set (template), and
6. The best match with one of these training characters is determined.

The line thinning and the feature extraction of this whole process have been mapped onto the chip. The minicomputer accomplishes all other operations.

In principle, the knowledge of a neural system is not prespecified by an algorithm or a set of facts and rules. Rather, information develops as a set of measurements made on various aspects of input patterns (as we saw in this character recognition example), as well as by means of network parameters that self-adjust during learning.

For instance, based on which label units are on after a run, a decision is made whether or not to delete pixels of the resulting pattern from the image. For the next run the pattern is shifted by one row or column of pixels on the image; the outcome is used as the input vector.

The number of stored vectors and their lengths is limited by the number of amplifiers in the circuit. Each component of the input vector uses one amplifier.

For each pixel position, a processing cycle is accomplished with the appropriate input vector. For each pixel in the image, a decision is made as to whether:

- That pixel just fattens the line, or
- Its presence is crucial to keeping intact the connectivity of the character.

Not only in character recognition, but tasks such as fingerprint analysis and inspection of manufactured parts line thinning are an important steps in machine vision. Many algorithms have been developed to handle this problem.

In America, in Europe, and in Japan, researchers in neural computing testing similar circuits with programmed algorithms as well as heuristics have concluded that such networks are *reliable* and *robust* enough for applications. Invariably they find that since input and output of data limit the processing chores, they must optimize the I/O structure.

Methods of statistical as well as structural pattern recognition can be mapped into neural network chips. Other artificial intelligence projects, for instance, on bit-mapped classifiers, contribute to the fine tuning of the neural network in building powerful recognition systems.

One of the promising applications areas in manufacturing is quality inspection. While the first generation of applications has given commendable results:

- In some cases variability is due to image-corrupting noise contributed by factors in the operating environment such as dust, vibrations, poorly controlled background illumination, and so on.
- In other cases, the variability is due to variations naturally occurring in part size, structure, quality characteristics, etc.

Contrary to the character recognition example we just followed, real-life part inspection tasks feature indicative quality aspects that are very difficult to describe formally and explicitly. A further challenge is the need to capture the fuzzy logic human inspectors learn to employ in making quality judgments.

In both identification and inspection tasks, for instance, intelligent vision systems must be capable of adapting to keep pace with rapidly evolving part specifications. They must also follow rigorous statistical inspection plans against which to gauge their output. The latter function can be nicely performed through operating characteristics (OC) curves.

Keeping in mind these fundamentals, the argument can be advanced that significant portions of identification and inspection tasks may be tractable with neural systems, mapping information processing strategies among networks of neurons. In this case, emphasis is placed on neural networks because of their ability to:

- Learn how to develop a set of experiences by training from examples, and to
- Follow a heuristic strategy in using these experiences to make correct judgments.

However, in spite of these projects and of reasonably good accomplishments, there are still important gaps in our understanding of neurocomputers. These include validation of the approaches they spontaneously acquire and, up to a point, associated computational limitations. There are, nevertheless, promising possibilities in this field.

3. THE CONTRIBUTION OF COGNITIVE SCIENCES

Whether formal or informal, knowledge is always based on the completion of something in accordance to a principle. In a scientific sense, this process becomes formalized into a *method,* that is, an established procedure whereby data flowing within a problem area is related to a principle that helps explain (or at least represents) incoming information in a comprehensive form.

A method is practically what we do when we arrange sales, inventory, or production data in a time series. At the same time, the way we develop the time series reflects the method that we use for representation:

- Aiming at a comprehensive picture,
- By adding an interpretation of trends and changes that must be watched.

Such trends and changes may reveal a pattern and its underlying characteristics. Perception followed by recognition are the fundamental steps in receiving a stimulus and in interpreting it; also in gaining insight in terms of possible orientation. In essence, fuzzy logic, discussed in Section 2, is *perception logic.*

The contribution of cognitive science is quite significant in the domain of second generation expert systems, and its usage helps differentiate them from the first generation of rule-based approaches that used brutal force. This statement is valid for fuzzy engineering and neural networks.

Whether we talk of human or machine implementation, perception is followed by *conception.* The mind patterns create coherence by recognizing schemes, developing frames, and making associations and concepts. The identification and recognition of patterns is guided by *metapatterns,* which are patterns about patterns, the way metaknowledge is knowledge about knowledge.

Intuition and memory often depend on a pattern and so do our different moves and actions. Said Judith Polgar,* "It was a home variant . . . I caught a view of the board and the mate pattern was familiar." How many patterns do we have in our mind? We do not really have an approximate answer to this question, but

- A wild guess is 2,000 to 50,000, and
- The lower figure corresponds to the number of Chinese ideograms.

Not only are we not sure about the number of patterns that exist in the human mind, but we do not know if these are prewired, programmed in a framelike manner, inherited, developed through training and experience, or done ad hoc.

What we know is that both with neural networks and with fuzzy logic we predominantly work in pattern recognition, emulating the way we think the human mind works. Some cognitive psychologists believe that pattern development looks like a realtime acquisition process, but:

- We really do not know how these patterns get fixed in the mind and for how long.
- All we know is that the process varies from person to person, and that training makes both the recognition of patterns easier and more effective.

Managing a pattern-oriented process is no easy feat since we do not really know how distributed and how dynamic these patterns are, nor what makes them dynamic. But we do know of their contributions:

- *In vision,* a pattern depicts motion, gets more comprehensive through color, and permits detail such as edge detection.
- *In memorization,* a pattern makes feasible the remembrance of faces, initials, names and the special skills of certain professional people.

There are reasons to believe that the animal brain was developed *by* and *for* patterns—from their conception to their interpretation and administration. The dynamics of looking around a pattern for interpretative reasons is a pattern in itself. Is there a kernel of pattern

---

* The 13-year-old who won the women's chess tournament in 1988 in 17 moves after sacrificing a queen and a knight.

methods? If yes, where does it lie? What are these methods? What are their advantages and their limitations?

Our ability to recognize patterns seems to be largely confined to a *scalar* view, but the full power of patterns shows itself in a *vector* space within an expanding universe. The transition from a scalar to a vector space is an elusive subject, but we do have some theories in other areas of pattern treatment. Within a particular line of reasoning, the pattern of the domain expert is based on:

- *Syndromes,* or events, items and concepts that run together,
- *Symptoms,* the observable phenomena associated with case descriptions,
- *Standards,* which can apply to professional metrics, and
- *Processing algorithms and heuristics* through searching, clustering, analyzing and studying taxonomy.

All this contributes to a professional person's emerging pattern regarding a given event or situation. His or her self-developed methodology includes values, viewpoints, contextual meaning, groupings, restructuring modifications, and hypermedia solutions.

Hypermedia approaches are navigational approaches involving information elements in the data base, but they also effectively explore the power of pattern recognition. This is what we try to do with object-oriented systems. A hypermedia solution has information stored in a network of nodes interconnected by links, with the whole creating a pattern.

- The nodes may contain different information structures such as text, data, graphs, images, voice or other objects.
- The links represent relationships between the nodes, and thus they contribute to the system structure.

Ad hoc structures made in response to dynamic queries are used as a navigational aid or browser. They underpin a query language and also the visualization tools a response requires. The navigational characteristics allow the user to browse through the hypermedia space by presenting him or her with a pattern.

Machine learning mechanisms too are pattern oriented or, more precisely, they are 90 percent pattern based and 10 percent logic based. And as Dr. Tibor Vamos was to comment, a similar mix is also relevant in human thinking: "Thinking is what we do when the pattern recognition fails."

4. PATTERN RECOGNITION AND OIL EXPLORATION

Pattern analysis characterizes many industrial efforts as well as a long list of research projects. For instance, in medicine this is true of diagnosing bacterial infections in order to prescribe therapy, investigate lung disorders, induce molecular structures from mass spectrometry data, as well as interpret the results of experiments in genetics.

A particular characteristic of this brief list is its close analogy to work done in oil engineering: The study of the results of multisensor integration, geological analysis, oil well log and seismic studies, and drill-bit problems is largely similar to the pattern analysis procedures that characterize medical applications. An implementation example is Baroid's MUDMAN.

Baroid invented drilling mud in the 1920s and it is now the largest mud company in the world. Mud is a lubricant needed in drilling oil wells. Baroid has a knowledgebase of nearly 70 years of mud experience, embedded in the skill of its mud experts who:

- Analyze rock formations based on samples,
- Create a pattern in their mind, and
- Reach conclusions based on it.

When Baroid sells drilling mud to an oil firm, it also assigns a mud engineer to stay at the site and to solve problems. It takes three to four years to train an engineer to think in rock patterns and to reach lubrication decisions, and no conventional algorithms are universally applicable to help in this work.

This is an ideal ground for the development of AI constructs. MUDMAN, Baroid's expert system, uses *pattern matching* and heuristics. It:

- Provides able assistance as an expert consultant,
- Serves the company's client base, and
- Expands the knowhow of the human experts.

Another expert system, the Norton Drilling Adviser, has been designed to assist directional well planners. It implements all knowledge in the Eastman Directional drilling manual and employs user defined limits on building and drop angles. It interpolates between survey points using cubic splines and it generates well reports showing dog leg, azimuth, direction, turn rate, build rate and formation.

Like the other expert systems we covered, the Drilling Adviser is interactive, and it displays spider plots (top view with multiple wells) as well as formations over a user defined arbitrary path such as an automatic interpolation between core samples. An associated visualization module handles multiple scales and user defined offsets.

There is a *dial tool kit* that simulates analog electric dials and pressure dials, handles compasses and vertical bar graphs with analog- and digital-style readouts, and supports multiple alarm levels:

- All dial types can be sized at the user's choice,
- The window system automatically scales dial components to match the specified overall size.

According to its developers, the Norton Adviser handles 80 percent of all possible drilling problems, providing advice and explaining reasoning strategy.

INTERISK is a model used to estimate the economic risk of developing petroleum reserves at the appraisal stage. This is particularly valuable when large capital expenditure precedes the start of production, as is the case with offshore drilling.

The financial analysis provided by INTERISK is based on production and cost profiles calculated by reservoir simulation, construction scheduling, and drill scheduling. The use of simulation accounts for the effects of uncertainties in:

- Reservoir data,
- Well data,

- Construction times, and
- Drill times.

It also permits for a number of variables to be evaluated. The model provides rapid and consistent economic evaluations of risk, indicates solutions that minimize risk, and compares the profitability of alternative development schemes.

One of the most successful expert systems in geology is *PROSPECTOR*. It is based on about 1,500 production rules linking evidence to hypotheses, but it also accounts for uncertainty in the evidence. The mathematics rests on Bayesian theory of conditional probabilities. The expert system:

- Uses a mixed initiative control strategy,
- Does backward chaining to gather information, select new goals, or input new data,
- Incorporates models of different kinds of ore deposits, and
- Expresses judgmental knowledge through a rule network.

PROSPECTOR is not a second generation expert system but its inclusion in this discussion is necessary to complement the other references. In its first version it was introduced in 1979. Since then, its semantic networks for expressing the meaning of propositions employed in rules have been improved and the same is true of the taxonomic network developed to represent static knowledge.

The late Jean Ribout, an investment banker and chief executive officer of Schlumberger (PROSPECTOR's developer), had this to say on the able use of new technology: "This technical revolution - artificial intelligence - is as important to our future as the surge in oil exploration. It will force us to design new tools, it will change the capabilities of our services, it will multiply the effectiveness of our instruments. It will change the order of magnitude ... of our business."*

Schlumberger invests heavily in AI, and most particularly in pattern recognition. This is the continuation of a tradition that started with the company's founder Emil Schlumberger, whose original discovery was a pattern analysis device that greatly assisted oil exploration.

5. PATTERN ANALYSIS AND METAKNOWLEDGE

Today, computer-based pattern analysis has basically the same aim that it had in Emil Schlumberger's time in the 1920s: To assist the human intellect in terms of perception and cognition, whether in a specific field such as oil exploration or in business at large.

While the tools of the trade have changed, with AI now making a major contribution, the fundamental notions and theoretical background are the same as they used to be. Cognitive psychologists maintain that great entrepreneurs combine in one person two quarter-spaces:

1. *Conceptual,* and hence inductive with ambiguity, the realspace domain.**
2. *Directional,* which is deductive and structured, the original field of realtime.

---

* Ken Auletta, *The Art of Corporate Succes—The Story of Schlumberger,* Penguin Books, New York, 1985.
** See D.N. Chorafas, *System Architecture and System Design,* McGraw-Hill, New York, 1989.

The advent of realspace increases the requirements for pattern recognition capabilities, and these can be effectively modeled by means of heuristics solutions.

Patterns can effectively assist both in conceptual and in directional domains, though the former profit from them more than the latter do. Formalisms are helpful. Although a large quantity of tabular data can be crowded into a small-time window, the result is not comprehensive. A pattern makes a complex relationship much clearer than data or words do.

Classification is a powerful patterning tool. To reach a decision and to take action, we have to understand *relationships*. This is as true of relationships linking the present, the past and the future, as it is of any other subject at any given point in time.

Three types of knowledge can be distinguished in information analysis, all of them helpful in classification and pattern construction:

- Declarative knowledge

This is a knowledge structure resembling a statement of facts, relations, objects, or events, rules, and heuristic mathematical expressions.

- Procedural knowledge

The construct is akin to an algorithmic structure and it is expressed as a sequence of commands for action(s). It allows for fewer degrees of freedom than declarative knowledge does, but it complements the latter in important ways.

- Metaknowledge

Metaknowledge is control knowledge. The construct contains knowledge about knowledge, telling how to use available knowhow and most importantly establishing constraints. Second generation AI constructs apply to the power of metaknowledge.

Most systems working with patterns *reason by analogy*. There are structural and symbol-oriented analogs, as well as constellations of events we treat through analogical reasoning. In fact, there can be a hierarchy of analogs as for instance in vision. Cognitive scientists usually discriminate among different levels of vision:

- Low level vision is identified with edge detection or similar elementary operations, focusing on linear metrics.
- A slightly higher level takes analogy as its way to proceed with the cognition of objects.
- The next level up is morphological analysis, building patterns of greater complexity from lower ones, analyzing, and then composing visual information.

We will return to the subject of vision in Chapter 8 when we talk of robotics. The reference made in this chapter both brings under perspective one of the foremost fields where pattern recognition is applied and it helps us to remember how important this branch of science is on the production floor.

Pattern analysis and metaknowledge underpin most developments in factory automation, but since new frontiers also present significant opportunities for error, we must provide ourselves with the means for fine tuning. This we do through prototyping.

Prototyping is a process that works through analogical reasoning. It is an application of our inclination to proceed through pattern structuring and pattern recognition. The building of models is synonymous with the development of an analogy; and the same is true of representations by worlds, situational logic, scenarios and mathematical expressions.

Finally, another contribution of pattern-oriented approaches and intelligent databases is *text animation.* Animation represents information contained in large chunks of text and data in the form of a knowledgebank. The knowledgebank is used to drive a computer-based consultation which permits the user to experiment and to provide for immediate visualization.

6. THE SEARCH FOR ENRICHED APPROACHES TO DATABASE USAGE

The search for new, enriched approaches to database usage definitely rests on the growing enduser requirements. Intelligent databases is a second generation example of AI implementation. With classical computers, efficiency is:

- 98 percent a function of database design, and
- Only 2 percent a function of the means by which the query is generated.

Idea databases, which is another way of saying intelligent databases, change this frame of reference. Query generation is done through AI and it carries by far the largest weight in the execution of the task.

Queries are indivisible from database administration, with the nature of queries having radically changed in the last few years, and it promises to do so even more in the the years to come. As Table 7-1 demonstrates:

- Queries used to be *crisp.* Crisp means true or false with no fuzzy sets.

TABLE 7-1. An Evolution in Man-Information Communication

Query	Database
1. Crisp*	Crisp and precise
2. Crisp	Crisp but imprecise (unknown, partial)
3. Crisp	Vague or stochastic
4. Vague**	Precise
5. Vague	Vague

* *Crisp* means true or false with no fuzzy sets.
** Queries are also increasingly ad hoc and can best be handled through fuzzy logic.

The problem with most programming languages today, including SQL, is that they have been designed for a crisp query environment—and for the handling of information elements in the database that are themselves crisp and rather precise. This situation however has radically changed; and Table 7-1 suggests that:

- Queries are increasingly vague and ad hoc, while the contents of the database become themselves vague, imprecise (though accurate), partial, stochastic—in short, *soft*.

Technology presents new types of software such as *query optimizers* to handle the new query requirements in an efficient manner. A query optimizer uses proprietary strategies to enhance the overall response capabilities, and to better the *response time*. Through *semantic modeling*, a query optimizer attempts to:

- Reduce the number of input/output or database accesses, and to
- Cut down the computation overhead for executing the query.

Most significantly, through the use of AI constructs, query optimization can be completely hidden from the endusers. This is one of the characteristics of intelligent databases.

As the foregoing references help document, we are moving towards a totally different database environment from the one to which we have been accustomed for more than three decades. New organizational forms are necessary and one of them is the concept of *episodic memory*.

In an episodic memory, multimedia information (text, data, moving image, graphics, voice) is stored in the form of object-oriented information elements—or *episodes*. These are not structured into files but can be automatically accessed through knowledgebank management systems (KBMS) as they constitute callable entities.

One of the first KBMS implementations, ICOT's Kaiser, is a distributed structure stored in:

- A myriad of workstations, and
- The large number of database servers attached to the network.

The aggregation of these parts constitutes what is known as the *distributed deductive database* (DDDB). The contribution of KBMS is particularly important, as the presently popular relational model has semantic limitations. The solution is semantic modeling.

Semantic modeling has become an important force in KBMS development. It is an attempt to represent the meaning of data so that:

- External schemata can be effectively expressed.
- Object-oriented solutions can be easier implemented.
- Appropriate integrity constraints may be maintained.
- Inferences can be made to aid user interaction.
- Ambiguity and fallacious associations can be avoided.

Representing the meaning of data in an able manner is the goal of *metadata*—, that is, of data about data. The use of metadata is a prerequisite to the able implementation of networkwide objects, which itself is a precondition to an episodic memory.

Coupled with very fast and reasonably low cost supercomputers,* adaptable database networks permit an exploitation of the advantages of episodic memory. As we know from current practice among the leading edge organizations, implementation of artificial intelligence in connection to the exploitation of databases follows one of three systems:

1. Rule-based systems
2. Domain knowledge, or
3. Episodic memory.

In the general case, rule-based systems exploit domain knowledge, but domain knowledge too has its solutions, one of them being the classification and taxonomy of which we have spoken in the preceding sections.

Episodic memory is event oriented, and it leads to a different type of reasoning: The intensive use of storage to recall specific episodes. This:

- Contrasts with the traditional assumption in AI that most expert knowledge is encoded in the form of rules, but it
- Makes feasible the *ad hoc* exploitation of multimedia databases.

As we saw in Table 7-1, old-style databases imply quantitative references and precise queries. In contrast, when using a knowledgebank, we only have an *idea* of what we *might* want and this is stated in qualitative rather than in quantitative terms. This is the wisdom of distinguishing between knowledge-type and data-type queries.

In other words, AI enriched data base systems differ from conventional computer programs by addressing themselves to data bases—and they do so in several important ways. One of the more visible is the way in which we make:

- *Knowledge*-type queries, and
- *Data*-type queries.

The difference is significant requiring a change in concepts if not in culture. This is a necessary ingredient of the transition from classical databases to idea databases and the use of knowledgebanks. At the start, such a transition may be confusing, yet it is part and parcel of the steady development taking place in information technology.

Let's say this in conclusion. In its fundamentals, technology is concerned with the systematic production of a class of artifacts. All specific efforts are characterized by two common features:

- *The need to be systematic.*

This means that there should exist an implicit and explicit body of knowledge that provides a conceptual framework, within which we can design and model new theories that lead to practical implementation.

- *A system of conventions* (and practices).

---

* The Connection machine, for instance, features 2,500 to 10,000 MIPS and costs about $3 million.

These are necessary to govern development processes and to provide standards. The latter reflects on the criteria for judging the performance and quality of artifacts.

By constructing elements from components and assembling them into aggregates of increasing complexity, we can reach specific objectives. This requires knowledgeable human resources as no goal can be obtained without them. It also calls for more powerful concepts and tools, which is precisely what we aim to do with second generation expert systems.

8

Advances in Robotics

1. INTRODUCTION

In August 1990 while the *Wall Street Journal* was headlining the public's rush out of emerging growth sectors, the *Harvard Business Review* was stating that young entrepreneurial ventures outclass larger competitors because: They develop products *twice* as fast, utilize assets *eight times* more productively, and respond to customers *ten times* faster. The new wave in robotics—from design to manufacturing—is part and parcel of this drive.

When robots first came into manufacturing, no company really had the storehouse of knowledge needed to run them effectively, but this has changed over time. During the last twenty years, robots have established themselves in industry primarily because of:

- A greatly improved productivity,
- An almost faultless performance, and
- The assistance in lowering labor costs.

Industrial sector after industrial sector have experienced the benefits to be derived from robotics, with Japanese manufacturers being among the world's most highly interested in their usage. They also derive the most benefits out of their implementation.

In a general sense, however, the word *robot* does not mean the same thing to all people. Its meaning varies as a function of culture, scope, time, level of technology, and extent of the implementation being done. At an earlier time, the definition of robots was limited to a machine performing one specific task that can be programmed (and reprogrammed) to do some other jobs. An alternative definition stressed the aspect of programmable multifunctional manipulators.

Earlier robot definitions did not include the use of artificial intelligence. Yet, today we start thinking of the ability to add intelligence to these machines in the form of vision, tactile sense, planning, and learning as tasks inseparable from the sense of a modern robot. We also appreciate that the crucial test in robotics implementations of the 1990s comes by way of how well an autonomous vehicle can perform without human guidance.

As we will see in this chapter, it takes much more than the old concept of robotics to master the manufacturing technology efficiently in the 1990s. There is a whole range of background studies that are necessary to support the next generation of robots in an able manner, including the ability to give them clear missions that do not contain too much carryover from the past:

- Modern production automation places a premium on efficient methods that require reasonable capital costs and low operating costs while sustaining high quality.
- Manufacturing engineering and product design built around the human component and dominating the production loop, no more qualify as a competitive edge environment.
- Advanced robotics solutions are using computer-based mathematical models such as algorithms, simulators and heuristics and integrating design, engineering and manufacturing for greater flexibility and productivity.

Intelligent robotics is what we are after: They can sense the environment in which they operate making logical decisions about how to proceed. Compared to them, the now traditional robots are often expensive, costly to program and of limited perspective, as they are useful for only one type of task, the goal being a flexible and intelligent usage of the best robots technology can offer.

2. TRENDS IN THE ROBOTIZATION OF INDUSTRY

Trends of a certain significance are developing in the robotics industry with regard to the technology being used, the user companies themselves, their type of applications, as well as the range of vendors. The importance of robots to industry at large is one thing; their specific impact on the competitiveness of some source of the industrial sectors and individual companies is another.

The emphasis on implementation is changing over time, and this shows in robot design. Ten years ago, everyone cited as examples the faster-than-expected developments in the area of assembly robots and the growing momentum for others such as arc-welding engines. Since then, competitive pressure forced the exploration of new markets and applications earlier than they might otherwise have been developed. This has been a dual approach:

- On one side is the competitive pressure for newer and better solutions including enrichment through microminiaturization and AI.
- On the other are the positive aftermaths in those manufacturing companies that dramatically raised productivity by leveraging technological leadership in such fields as robotics, computer-aided design and materials science.

During the mid- to late-1980s, the automakers of the First World had plenty of capital with which to finance high technology retooling. The cash flow and earnings quality of capital-intensive companies showed an impressive growth with the Reagan recovery, thanks to low capital goods inflation and more favorable treatment of depreciation.

Additionally, under fire from foreign competition, labor and management learned to work together to maximize total profits instead of squabbling over the question of how large a share of the pie each would get. With the new spirit in manufacturing, the goal became the opposite of Henry Ford's Model-T.

An example is given by the Fiat engine factory at Termoli (on the Adriatic coast) which, when it was opened in March 1985, was a technological breakthrough. Built to produce a new engine, the Fully Integrated Robotized Engine (FIRE 1000), for small cars it marked the first successful attempt by an automobile manufacturer to:

- Design and develop *a new factory* and *a new engine* together at the same time.
- Take an *integrative* approach to shape each one of them to work best for the other.

As an automotive manufacturer, Fiat designed the technology to produce this engine and developed the engine to be produced by this technology. Six years later the Termoli engine plant is still one of the most highly robotized plants in the world.

Another major innovation in manufaturing has been a fully robotized body and final assembly plant at Cassino (near Naples) where Fiat's new medium-size model, the Tipo, went into production in early 1988. All the elements of the car from engines and gearboxes to the doors and seats, are assembled and fitted into the body by a fully automated system of computer-guided robots.

One of the basic characteristics of the Cassino robotized factory has been its flexibility. Its design foresees that the process of final assembly may involve as many as 5,000 variations, depending on:

- Styling,
- Options, and
- Finishing touches.

Such variations are necessary responses to constantly changing consumer tastes. This approach goes a long way from the concepts of the late 1970s when paint spraying and spot welding were the main jobs robotized.

This is precisely why with the new trend in automotive manufacturing the goal is the opposite of Henry Ford's Model-T. Instead of the largest possible number of cars produced by the same machinery, the world's foremost auto manufacturers are coming up with a technology that:

- Permits them to change in the shortest possible time at the lowest cost
- From one variation to another, and from one model to another.

The goal is not to produce the greatest number of cars most cheaply but to get the lowest possible break-even point at which production of *small lots* representing the variation of a model can be held and still *be profitable*.

In other fields of activity, too, companies are making major commitments to the use of industrial robots in a flexible way. This is particularly true of firms today that are about to launch a sweeping automation program that may eventually replace half of their assembly workers with robots. GE already figures that it can eliminate a four-digit number of blue-collar positions in relatively short order, capitalizing on the fact that:

- New technology is making it possible to have robots considerably more versatile than their simple-minded predecessors, and
- The generation of robots that *see, feel* and *think* is not only steadily emerging but also their development process is accelerating.

In a similar vein, the Japanese are installing new robots not simply to automate but also to make production lines more flexible. Nissan's newer auto plants can produce hundreds of different variations on a given car model simply by reprogramming robots that:

- Go beyond painting auto bodies,
- Install car seats, engines, batteries, windshields, tires and doors.

This is one more example of Japan's skill at grasping an evolving technology and putting it to work. This includes its thrust for *metal employees*. It has happened already in consumer electronics, memory chip production and machine tools.

As we will see in this chapter, fixed automation is of the past. The competitive edge is AI-enriched automation and, with it, autonomous vehicles. At present we are in a delicate phase of transition; the old robotics is dying out while the emerging robotics solutions will have a great impact on our future industrial and cultural tradition.

3. ARTIFICIAL INTELLIGENCE IN ROBOT CONTROL

Robotics is a good candidate for the application of AI technology since it includes many aspects of human reasoning and a range of motor skills. Adding intelligence to the metal employee permits robot vendors to market their product into a wider variety of industries, as well as to adopt a more aggressive marketing posture.

Since the market for robots has always been quite concentrated in certain industries and in specific companies, a broader sales approach carries with it the challenge of educating the user organizations as well as increasing product visibility:

- Visibility is improved through breakthroughs in autonomous robots, and
- Autonomous robots need artificial intelligence from vision to the ability to learn from experience. They also need to follow voice commands.

Vision and foresensing systems are available today. More impressive is a development that could add new dimensions to robot adaptability, for instance, programming with AI so that a robot facing an obstacle would try different solutions, learning how to overcome constraints in its environment in order to continue its task.

Research data tend to support the hypotheses that mobility, acute vision and the ability to carry out survival-related goals provide a necessary basis for the development of true intelligence. AI research typically produces programs able to handle important domains of intellectual activity:

- Cognition,
- Spatial understanding,
- Sensor input,
- Noise filtering, and so on.

Using advanced robotics and artificial intelligence, a US Army scout car was computer driven from a remotely located command post over a mile away. During this first-of-its-kind demonstration, the Advanced Ground Vehicle Technology program utilized two subsystems:

- Autonomous Vision transmitting video images of the road to computers which sent back steering, brake, and throttle commands.
- An AI-based Map And Planning scheme which kept track of the vehicle at all times, displaying its location on a map of the local area on a color monitor.

The system was operated day and night with thermal sensors and a complex communication link that coordinated the overall system function.

AI, computer vision and touch-sensitive arms make robots more complex. But they also make them much more flexible. Flexibility and adaptability are among the major themes running through the robotics industry, with companies showing new robots in which tools on the end of an arm can automatically be changed so that multiple tasks can be performed unattended.

In a manner quite similar to that of the generation of supercomputers now coming into the market, the emphasis is on providing *solutions* to manufacturing problems. This is a far cry from past policies that simply demonstrated a given robot, leaving it up to the user organization to figure out how best to employ it.

Basic to this solution-oriented strategy is the concept of *workspace definition* which contains:

- Problem description, and
- Problem status.

Communications channels and the knowledgebank provide for domain knowledge through rules, frames or scenarios. AI assures concurrency of access to databases and thus it supports distributed operations. It also provides for the resolution of inconsistent information.

Residing in the knowledgebank, the constraint mechanism is a good way to *specify* and *supervise* relations, for instance, flashing out inconsistencies in assumptions and/or perceived patterns.

Information processing in the robot will be increasingly pattern-oriented, based on a distributed, parallel, modular approach instead of the central locus of the control approach typically followed in many traditional robot applications. The new approach is leading to a natural modularity and hence to greater robustness.

What is more, perception and motor behavior starts being integrated with knowledge representation. The robot system's intelligence becomes adaptive: It learns simple spatial relations and it is able to recreate them.

While many of the skills that make up intelligence are used even by lower life forms, others such as language, abstract symbol construction, and formal manipulation are employed extensively (as far as we know) only by humans and the most sophisticated manmade artifacts:

- Through modeling and reasoning, robots equipped with AI constructs can assimilate differences between observation and expectation.
- According to cognitive scientists, structural features are more important than accurate dimensions in deciding object statuses.

For object recognition, robot sensors provide information on objects to be recognized by analyzing the observed data using a priori knowledge.

For example, using object recognition and status identification techniques, a computer system for multimodal matching (which has been shown to play a vital role in human information processing), has been developed for autonomous mobile robots. The matching processes collaborate by sharing the same model, thereby improving the efficiency and reliability of mobile robots, which can then navigate themselves at a practical speed.

Knowledge-representation structures employed by a robot's AI construct, in conjunction with sensory devices, form models that can subsequently be manipulated. Such models may reflect the robot's external environment as well as the robot's own physical configuration and internal state.

To be employed in an able manner, such representation must be complete, allowing all important aspects of the situation to be permanently represented. It must also be adaptable so that the stored information is used efficienctly. An efficient structure is one that is matched with available processing capabilities, anticipating uses to face ad hoc situations. Only then can we talk of intelligent robots.

Knowledge-based developments are becoming increasingly crucial in leading-edge industrial and business applications. As the robotics industry steadily refocuses, emphasis is placed on selected areas including new or upgraded control devices as well as a growing pallet of robot capabilities. An important part of the latter reference is the combined realtime use of computer-based sensor and thinking technologies.

4. TOWARDS AUTONOMOUS VEHICLES AND MICRO-MINIATURIZATION

The concept of an autonomous vehicle is a dream of automation designers, but it is not all that new. Called *rolling robots* the first species involved a modified, battery-powered tow truck that followed an overhead guide wire strung inside a grocery warehouse. This happened nearly 40 years ago, back in 1953. Two decades later, in about 1973, Volvo (of Sweden) adapted the concept to a manufacturing setting.

Primitive as they might have been by today's standards, this first species was a significant development as it opened new perspectives. Today, a large population of automatic guided vehicle systems (AGVS) from a variety of manufacturers are operating along this line. Rolling robots have wheels instead of arms and handle the in-plant transport. They come at various degrees of sophistication:

- The not-so-complex are like pallet trucks with a sort of forklift.
- Others are tractors that can pull trailer trains.
- A more versatile species is the platform carrier on a unit-load transporter (ULT).
- The most sophisticated combine lessons learned from the past with microminiaturization and artificial intelligence.

This latest generation of microminiaturized robots is a far cry from the clumsy automatons with hooked arms; they are little things the size of silicon chips with legs or wheels. Their designers see millions of them rousting barnacles off ships, repairing fiber optic cables, or exploring distant planets.

If this technology pans out, such minisize (palm-held) autonomous vehicles could be quickly produced and launched in a myriad of jobs. They are fast and cheap with a reasonable amount of intelligence—a truly new generation of mobile robots.

Micromachines were pioneered mainly at Bell Laboratories and at leading universities including Massachusetts Institute of Technology and the University of California at Berkeley. The earliest work involved using semiconductor manufacturing techniques to craft tiny gears, sensors and other parts onto silicon chips. Already, microlithographers can etch motors onto chips.

In the microworld, however, there are many curious phenomena. The tiniest obstacle is enough to stop a moving part, and because of surface tension, lubricant materials no longer lubricate, and they tend to act as adhesives. Yet, Japanese researchers have been eagerly developing this technology all along:

- Toyota Motor Corp.'s Central Research Lab has a prototype of a micropressure sensor to be used in its electronic fuel-injection system, and
- Tokyo University's Institute of Industrial Sciences is building superconductive micromotors to solve friction problems in chipmaking.

While a lot of challenges have still to be sorted out in the world of miniature autonomous vehicles, the lesser sophisticated earlier species of moving robots have found productive jobs to do. The ULT, for instance, can hold up to ten tons and double as tractors. The more complex among them have built-on lift/lower tables to assist at workstations on board, ready for action wherever they are sent.

Most important are the acceptance rolling robots found in industry, particularly within an integrated factory automation setting. For automakers, Toyota City has come to mean a manufacturing system that:

- Clusters parts plants around assembly plants,
- Provides for speedy delivery of parts and just-in-time inventories,
- Practically eliminates expensive operations by building cars with less labor.

All this rests on the implementation of transport robots, of which the rolling clan is an integral part. In a number of cases, this reworking of production facilities resulted in technically advanced manufacturing and assembly operations.

For all practical purposes, therefore, robots are moving out of the protected drudgery of factory assembly work and into the manufacturing environment at large. A number of approaches also aim to introduce metal employees to nonmanufacturing jobs as coal miners, road crews, and construction workers.

This broader implementation trend sharply contrasts with the first generations of robotic machinery which was mostly stationary, restricted to repetitive manufacturing operations. Making robots mobile requires:

- Rugged mechanisms,
- More complex technology, and
- Sensory systems for avoiding obstacles.

Sheltering the robot becomes a problem and it may be years before robots are used widely outside the controlled environment of a factory, and are given the ability to move about independently.

Radical changes in machine design demand new departures. Rather than trying to robotize a conventional piece of machinery, such as an earth-moving traction with sensors and computer intelligence, it is more rational to reevaluate tasks now performed by conventional machinery, designing robots specifically for those tasks in a way similar to the new factory and new engine example of Fiat.

Radical redesign calls for more imaginative solutions to office and manufacturing environments, leading toward new technical solutions that constitute a factory aggregate. Today, we do not yet have the technology to built such complex system aggregates but we have the ability to move toward them as we will see in the next section.

5. THE ENLARGING DOMAIN OF FIELD ROBOTS

Field robots must contend with a highly unstructured environment. That means it is not possible to program the robot's computer for all possible contingencies, doing so in a procedural manner. This contrasts with factory robots which are typically bolted to the floor and need only repeat the same set of motions time and again.

If a field robot is to home in on its mission or avoid falling off a ramp, it needs:

- Artificial senses, especially vision,
- Fast responses, and
- The intelligence to make independent decisions.

A decade ago, seeing an obstacle and identifying it required minutes of processing time by the robot. Today, it takes mere seconds; this is about the best performance achieved by researchers working on the Defense Advanced Research Projects Agency's (DARPA) autonomous land vehicle program.

The interest of the military in field robots is understandable as the US Army has estimated that about 50 percent of the Earth's land surface is inaccessible to all conventional vehicles, whether they run on wheels or on tracks. This is one of the reasons the armed forces is interested in the ASV, a vehicle that can walk anywhere.

Financed by DARPA, ASV is built for rough field use, carrying its own motor and accessories. The vehicle's top speed will be only 8 miles (13 kilometers) an hour, but it can go on terrains that a tank or an off-road vehicle would bog down; and it can be adapted to move and to fight on its own, without a human driver or crew.

This is only one example out of many. Field robots are typically equipped with wheels, tracks, or legs, and they have started to perform such tasks as cleaning up hazardous-waste sites, putting out fires, disposing terrorist bombs, and spotting drug smugglers. The following is a list of their implementation domains:

- Assessment, analysis, sampling, excavation, and cleanup of hazardous-materials sites.
- Inspection, repair, and cleanup of commercial nuclear power plants.
- Military reconnaissance, materials handling, and runway repairs.
- Bomb detection and disposal and chemical accident cleanups.
- Sea research including rig inspection for the oil industry.
- Space station construction and maintenance and exploration of other planets.

What most of the moving robots have in common is the need for a very careful planning for geometry which is more sophisticated than other applications of AI. The planning algorithms currently used in AI are not sufficient. The moving robot needs tolerance information in its program.

Moving robots require both sensors for balance and a way to assure that this is effectively done. The simplest way to balance is produce a hopping machine. This requires two strategies:

- *Control position* while in the air which means predicting where the foot will land;
- *Control torque* so that when it lands, body inertia will play a role and torque the body.

Autonomous vehicle technology often uses three controllers: forward-running velocity; hopping height; and body attitude. In terms of implementation, lessons are learned from the animal kingdom. The kangaroo for instance, uses less energy when running than when walking; and the role of its queue is not only to balance but also to store energy.

One of the developments has virtually been a leg machine, combining hopping machine characteristics in one leg. A two-legged machine that does not hop but walks has been developed for the American military and it has moved with a speed of up to 20 mph.

Research projects focusing on the development of four-legged machines incorporated principles found in the:

- Hopping machine (a one-leg mechanism), the
- Leg coordination (a horse-type strategy), and the
- Software for walking and running.

Software, and most particularly sophisticated software, is an integral part of any autonomous vehicle mechanism. Languages are the crystallization of all work done so far in robotics.

But the references made to so many different projects of autonomous vehicles should not be confused with the number of walking robots used in daily work. Just a handful of walking or rolling robots have been commercially employed since the late 1970s and until the mid 1980s, most of them were used for underwater maintenance of offshore oil rigs. Since the mid-1980s, however, there have been hundreds of them around and they could soon number in the thousands, including some in orbit like astrobots ready to repair faulty satellites.

Martin Marietta Corp., for example, is designing a prototype for assembling and servicing the planned US space station. NASA projects too will send this robot on a shuttle flight in 1992, for an in-orbit test.

6. PRACTICAL APPLICATIONS OF MOBILE ROBOTS

Construction is a good potential application of mobile robots because of the hazards involved, particularly high in the air, below ground, and deep under water. With seed money from the Japanese Ministry of Construction, large contractors such as Shimizu, Taisei, and Kajima have built mobile robots for spraying fireproofing materials on columns and beams, pouring concrete into form work, and finishing the surface of poured concrete.

In every one of these implementation domains, robots can improve productivity and cut labor costs, the more so since average wages in construction are considerably higher than those in plant manufacturing are. Other possible applications include autonomous robots capable of inspecting and maintaining large suspension bridges.

In contrast to the miniaturized robots of which we spoke in the preceding section, an eight-legged, 70-ton robot manufactured by the Komatsu Corporation in Japan has replaced 50 human divers in the construction of an underwater foundation for a seawall. This robot, which measures 56 by 33 feet (about 19 by 11 meters), uses its eight legs to make its way through water up to a depth of 90 feet (30 meters).

An eight-legged structure, the Komatsu robot, anchors itself on four legs at a time, moving the other four legs forward. Construction workers have successfully employed it in different trial projects.

Toshiba introduced a robot designed to climb ladders. The 3-foot 4-inch (about 1 meter) machine can carry a 15-pound load up a ladder at a pace of 13 feet (more than 4 meters) a minute, using photoelectric eyes to locate each step. It then uses small electric motors to position one of four grippers around the rung and it shuts the clap.

Workers, for example, can use the robot to shuttle equipment in nuclear power plants and other dangerous work environments. Made by Odetics, another robot designed for work inside nuclear power plants is targeted for spaces where there mazes of pipes exist.

Odetics's Intelligent Machines Division also envisions a 30-foot (10 meter) device, made up of 2-foot segments with wheels on each side, that will carry tools on its back.

Many intelligent autonomous vehicles have been born by necessity—which is always the mother of invention. Pennsylvania's Three Mile Island (TMI) nuclear accident in 1979 proved the viability of service robots and prompted Bechtel Corporation to find a cost-efficient way to carry out its $1 billion, nine-year cleanup contract.

Decontaminating TMI's No. 2 reactor would have involved 21,600 human *jumps*.* So Bechtel paid $800,000 to the Robotics Institute of Carnegie Mellon University for three robots, including a six-wheeled model that could carry 1,000 pounds of equipment to clean or to break down contaminated walls. This engine also removed 100,000 gallons of radioactive sludge, all in all an outstanding job.

At each of the power plants serviced by Westinghouse, a 77-inch-long robot, known as Rosa, inspects the nuclear reactor. In one hour it can control the pipes making more and better welding repairs than squads of human jumpers can.

Since Westinghouse put the first Rosa model to work in the early 1980s, it has slashed power plant downtime. This is saving utility-company customers at least $500,000 each time the robots finish a chore one day sooner than humans do. In the US other efforts along this line include machines that build brick walls, line tunnels, and replace worn shovel blades on strip-mining machinery.

Next to immediate necessity, a driving force towards intelligent autonomous vehicles is long-range planning. During the mid-1980s, the Japanese government launched together with Japanese firms a seven-year program to develop mobile robots with humanlike senses. This $100-million MITI project is nearing its end, having produced prototypes for three jobs:

- To clean up nuclear power plants,
- To fighting fires, and
- To assist in undersea construction.

During the project's life cycle, development tasks were parceled out to 20 companies. Computer makers Toshiba, Fujitsu, and NEC concentrated on *vision* systems. Fuji Electric developed manipulators possessing both strength and a delicate sense of *touch*. Ishikawajima-Harima Heavy Industries crafted a metal skin for the firefighting robot.

All mobile robots have been enriched with AI. The Ishikawajima construct, for example, has artificial "sweat glands" that let it withstand temperatures up to 1,200 C. Fanuc built especially powerful but also smart actuators for arms and legs. Given the merger of robot technology and knowledge engineering, the species of this focused effort will be felt in the world's markets and manufacturing processes before too long.

---

* The term comes from the usage of human *jumpers* in the nuclear industry. These are technicians trained for and assigned to work in dangerous atomic factory jobs. They must rush into the area and rush out of it in forty-five seconds.

7. ARTIFICIAL VISION AND FUZZY ENGINEERING

In Chapter 7 we said that artificial vision, computer vision, machine vision and AI vision are practically synonymous terms. They describe the technology of analyzing and identifying the contents of a scene from images of that scene. This has many major applications in areas such as:

1. *Industrial robotics,* including automated visual inspection, quality control tasks, and vision for robot guidance.
2. *Document processing,* including character recognition, signature recognition, and computer understanding of drawings and maps.
3. *Medicine,* including automated blood cell counting, tumor detection on radiographs, other diagnostic operations, and
4. An *expanding horizon* of implementation areas including remote sensing, cartography, reconnaissance, and so on.

The implementation examples of mobile robots and other constructs that we have seen in the preceding sections help to demonstrate how critical artificial vision is. By the same token, this reference underlines the fact that AI is a kernel subject in the development and use of robots.

Both algorithms and heuristics have been developed over the past decades for computer analysis of images. As the power of available computers continues to increase, and does so at an affordable cost, more and more of these models are finding practical applications, provided there is foresight as to how they can be used.

Computer vision does not need to be done only to emulate the human eye. The most successful projects are those that use both abstraction and imagination, starting with perception and cognition. Classification work—a major area in AI implementation—plays a major role in this process. One of the fields where research and implementation have obtained concrete results is pattern recognition.

- In AI, *pattern recognition* refers to the ability of a man-made system to recognize and classify patterns.

Typically, such a construct will take a data input, identify and classify its contents, develop trends and correlations, and find an associated output which is a pattern.

- In contrast, *pattern reconstruction* refers to the ability of a man-made system to reconstruct an incomplete pattern.

Starting with such an incomplete pattern, an AI construct is able to insert the missing information by retrieving the best complete pattern associated with the input, for instance, through the use of templates.

Second generation expert systems, such as neural networks, can be trained to identify and complete partial schemata or to distinguish trading patterns that are incomplete and fuzzy, restructuring them by using information they find in a distributed data base as well as in oncoming data streams from information providers and other sources.

Some of the results obtained through these solutions are impressive. Yet, as the preceding sections repeatedly stated, while technology keeps evolving quite rapidly, in industry we are faced with the monumental task of applying it in a practical manner, solving real manufacturing line problems. In this sense, our most common mistake is underestimating how much work and how much skill it takes to turn a theoretical breakthrough into a practical application.

Today most of the work done on pattern recognition deals with objects we consider as having some standard pattern. But the size and shape of natural objects (such as plants, animals, market trends) vary tremendously and never have a true normalized reference to be taken as a *standard*. Something similar can be said of a number of factory operations.

Since pattern recognition in processing operations is a new branch of science and experience with it is still thin, we will be well advised to learn from any domain we can—including its tools and methods—after adapting them in manufacturing activities. Dr. T. Terano, S. Masui, S. Kono and K. Yamamoto have studied pattern sizes and shapes.*
The researchers:

- Represented their features by means of natural language, and
- Processed their patterns by using fuzzy logic.**

The approach taken by this project rests on search trees and the researchers have demonstrated that once the hierarchical structure of feature items in a search tree is known, irregular shapes can be efficiently recognized.

By analogy to R+D work as well as to the operation of production facilities and market trends, the recognition of crops through AI makes an interesting case study. Pattern recognition through fuzzy engineering capitalizes on the fact that no two patterns of size, shape, location and direction are the same. A dual problem must be solved in order to proceed along this path. We must:

1. Find what the special character of the object being perceived is, and
2. Establish a methodology to represent this special character in a form appropriate for machine cognition.

Professor Terano and his coresearchers applied Principal Component Analysis (PCA) to the No. 1 issue and fuzzy set theory to the No. 2 issue. In the study of cognition, they used natural language as the mediator. This work is most commendable; it is also a theoretical breakthrough that can lead to the solution of many challenging problems in artificial vision.

---

* From a personal meeting with Dr. Terano at Hosei University, Kajino-cho, Konganei, Tokyo, Japan.
** It is not the goal of this text to discuss fuzzy logic. Though we have seen its underlying notions in Chapter 7, for more background the reader may wish to refer to D.N. Chorafas, *Knowledge Engineering*, Van Nostrand Reinhold, New York, 1991.

8. HIGH-LEVEL VISION SYSTEMS AND SIXTH GENERATION COMPUTERS

To fit properly a theoretical solution within a practical real-life environment is more or less the same as being successful with robot applications in a flexible automation sense. This requires that we understand the motivation of robot vision, its mechanics, and the need for notions more advanced than those we have had so far:

- Visual sensing and image segmentation techniques must be defined, not just described.
- Shape description and recognition methods, the interpretation of 3-D scenes, and shape analysis techniques must also be brought into perspective.

Vision systems are not monolithic: Both their sophistication and their flexibility are on the increase. The so-called *low-level vision systems* are not knowledge based in the current sense of the term. They are not equipped with rules so that they can be used flexibly in unanticipated ways.

But low-level vision systems do measure various features of images and, in some cases, they can be efficiently implemented with robotic hardware. They may also constitute the lower layer of a more sophisticated solution, and hence, a crucial part of the overall approach.

High-level vision systems employ AI reasoning techniques and are able to cope with ambiguities. They can handle some ad hoc situations and can also expand in other directions.

Like vision, *touch sensing* is a kind of image processing; and again like vision, it makes use of image enhancement, analysis, and identification. Hence, there are conceptual similarities between computer vision and artificial touch sensing capabilities.

There exist two- and three-dimensional (2-D, 3-D) computer vision systems. Both prove to be relevant to most manufacturing and assembly processes, even if some of them are just beginning to work well on the factory floor. For instance, sensing technology uses visual data from television cameras to:

- Help in quality inspection,
- Identify parts, and
- Increase guidance or control in the manufacturing process.

An implementation area of great importance for computer vision is in fact *inspection* and *quality control* (QC). Today, more money in industry is being spent on inspection than on manufacturing, and artificial vision is the key to the automation of the inspection process, both improving the outgoing quality of the products and swamping costs.

Here we are talking about wide-ranging benefits to be derived from scientific developments if we are careful to use technology in a pragmatic manner, within well-established timetables, and with due appreciation to the fact that AI systems work best in demanding environments. Typically, such environments are:

- Rich in data,

- Subject to many types of noise,
- More or less prone to error,
- Computationally very intensive, and
- Work online to other applications.

Although some of these environments are more classical and others are much more advanced, in both cases, understanding the AI principles that can help the application is not only a scientifically interesting issue, but it is also critical for assessing the solution's utility, flexibility and future potential.

However, the problems embedded in these processes are far from being solved. A successful laboratory demonstration of new computer vision techniques only means we are halfway through. The burden of working out the application often rests with the manufacturing engineer and the plant manager. They should be assisted by knowledge engineers:

- Management, who only cares about *buying technology,* is not very interested in knowing the tedious details of how to make this technology work properly.
- But open-ended management realizes that *applications engineering* is the vital link as well as a basic responsibility, even if it is not often appreciated as being so.

In other words, while the mathematics of computer vision is an exciting scientific subject and a prime field of AI research, even the most imaginative solutions will be of no avail unless they are enriched with applications perspectives. There is a world of difference between developing computer vision and implementing it. That is what many start-up companies with brilliant AI research people did not understand, and they went under.

Knowledge engineers are needed on the implementation side because directly interacting with the real world is a primary challenge. The correct answer to practical questions can make the difference between the success and the failure of robotics projects. In industry as everywhere else, success is what happens when *preparation* meets *opportunity*.

A similar statement is valid regarding the now-developing new species of *sixth generation computers* (6GC) that will be characterized by four milestones we are approaching at a reasonably fast pace:

1. Machines that are equal to man in intellect.

That is, *android brains* to be employed in increasingly sophisticated and hazardous tasks as we have seen in the preceding sections of this chapter.

2. An effective combination of machine intellect and electromechanical power.

If we produce *autonomous vehicles,* then we prove 6GC. This is a subject which has recently attracted a growing amount of attention.

3. Intelligent, wideband communication networks.

This is the goal of the Japanese *Advanced Telecommunications* Research (ATR) among its other projects. It is as well a milestone of community intelligence.

4. A great power of abstraction.

To conceptualize the advanced systems that we now project, we must go beyond the system now available working at the *meta* level with the lower layers inheriting the characteristics of the higher ones and the latter being in control of the former.

Presently, in advanced research and development status, sixth generation computers will feature the ability to: Combine machine intelligence and the human brain effectively; develop self-learning capabilities supported by AI constructs; and incorporate artificial perception and cognition, as well as artificial vision.

Fundamental to 6GC development is the integration of multiple channels of stimuli. This calls for solutions that go beyond logic, requiring metalogic for conceptual modeling. It also makes advisable:

- At macrolevel, multimedia object-oriented communication networks, and
- At microlevel, autonomous vehicles with self-controlled motor activity and AI.

Quite clearly, this type of implementation will take years to materialize, but by the end of this decade it will be within reach if is not already reached at least at the prototype level.

The competitive edge will not come as a matter of course to any company and to any nation. Attention to detail has many implications in management, manufacturing and engineering—all the way from the production floor to industrial policy.

It is high time to move away from macrostructures, previously used for their economy of scale, and toward microstructures. With microstructures, AI becomes the brain-work operator. When we exercise insight and foresight, high technology acts not only as the agent of change in industry but also as a prized flywheel bringing security to a process of transition.

part 2

COMPUTER-AIDED DESIGN, PRODUCTION SCHEDULING, QUALITY MANAGEMENT, MARKETING, AND COMPUTER-INTEGRATED MANUFACTURING

9

A New Generation of CAD Solutions

1. INTRODUCTION

In the 1970s came CAD and though the term stood for computer-aided design, it was mostly applied for computer-assisted drafting reasons, that is, at a much lower level of sophistication. Early CAD approaches focused on using a computer to do the rather mechanical but time-consuming drafting job and not to generate the deatiled drawings needed to turn engineering specs into usable engines.

Over the years, CAD practically eliminated drafting in the manufacturing sector. Then, as CAD users became more experienced, they started focusing on design proper, integrating it with engineering and manufacturing applications that had been supported through computers. An example of this is the production of the Bill of Materials (BOM).

As the sophistication of CAD approaches increased, so did the benefits to be derived from them. Not only did design engineers working with CAD tremendously improve their productivity* but they also increasingly used artificial intelligence enriched tools to obtain:

1. Better integration among applications,
2. More generic product design approaches,
3. Better fitness to do the work at hand,
4. More advanced implementation perspectives, but also the
5. Means enabling them to integrate past drawings that had been kept on paper as aperture cards.

Companies and people with experience in CAD do appreciate reference No. 5. In the late 1970s, I helped install CAD in a leading manufacturing company which, in order to integrate

---

* See also D.N. Chorafas, *Engineering Productivity through CAD/CAM,* Butterworths, London, 1987.

past and future engineering work (on paper and CAD respectively) had to bring into the engineering data base 1,250,000 drawings of all types. The sheer size of the job made it impossible to do without using AI approaches.

The need for advanced solutions is all over. For instance, along with reference No. 4, MIT currently experiments with a method of computer-generating the molds used to manufacture metal parts for autos or aircraft engines:

- Traditionally, molds are carved from wax, dipped in a ceramic mud, and left to harden. Then the wax is melted from the shell.
- MIT researchers say that their *three-dimensional printing* process cuts such mold-making techniques from six months down to a few days.

A computer-controlled nozzle applies a liquid adhesive spray, thin layer by thin layer, until a ceramic mold up to three inches per side is formed. Unlike other mold-making techniques that employ plastics or resins, the MIT process produces a mold that directly accepts molten metals.

Besides working to make more detailed molds, the researchers are attempting to mold finished parts by spraying metal powders as well as ceramics. If in the end this process proves to be truly successful, mold printing might just do for manufacturers what CAD did for designers—further document high technology's significant contributions.

The metareference carried by this example should not be lost: We have to apply innovations steadily across the board of the manufacturing processes. CAD practices can be enriched not just at the conceptual level but also through contributions made by means of advanced implementations that eventually feed back into the original design.

2. TOWARDS SECOND GENERATION CAD

Developments such as those we considered in the introduction are interesting in that they trigger a steady evolutionary process connected to CAD applications. This is in fact the role of expert systems developed to assist engineering design at the initial, middle, and final stages.

Second generation CAD is instrumental in correcting manufacturing industry's biggest disadvantage: its slowness. Detroit, suggests *Business Week,** replaces old models an average of every eight years. Ford averages 9.5 years; most Japanese carmakers average 4.5 years, but Honda averages 4.

That lead time gives the Japanese more chances to improve styling and technical innovation as well as to adjust to surprises. For instance, as fuel economy becomes a big selling point because of the on-and-off oil crisis, Japan's more flexible manufacturing techniques allow it to reconfigure its product mix faster than Detroit can.

No wonder that American auto manufacturers are now engaged in imaginative long-term projects to rethink the way cars are:

---

* October 22, 1990.

- Designed,
- Fabricated, and
- Sold.

As in mathematical research, they exercise logical schemes to establish how well the whole design, manufacturing, and sales process are consistent and to observe what expected and unexpected phenomena they contain.

Surprises can be found all over. Robot logic, for example, is intricate, and the sense of surprise and discovery of the unknown is present here as it is in laboratoty research.

Powerful solutions result from testing and refining the design. Recent approaches aim at arriving at an expert system formulation that captures and explains a set of real world phenomena. This calls for the same rigorous approach used elsewhere in science, and it has similar pitfalls, and is equally rewarding.

In one application, for example, an expert system supports lens type design in the initial study phase:

- If the designer specifies changes in lens system specifications or in the value of a basic parameter,
- Then the AI construct automatically generates a new lens solution using CAD as well as smart methods of error recovery.

Among companies in the precision machinery industry, Citizen Watch has developed AI-enriched approaches to integrated circuit design, numerical control and machine tool design. Another company built into its engineering design software a fault-diagnosis expert system for copy machines. All this is a far cry from where CAD stood at its beginning, only some 15 years ago, and this talks volumes of the great strides that have been made in its domain.

A large number of expert systems have been used in engineering design and development. In Japan alone, forty-one expert systems have been reported, including those for the VLSI circuit design, the automated engineering design of a power switching source, the design of nuclear power plant equipment, computer room layouts, and die-processing design support.

Computer-aided manufacturing (CAM) has also benefitted from this trend towards AI-enriched solutions. In production management alone, a golden horde of expert systems have been reported, including areas such as fault-diagnosis for ion injection equipment of semiconductor manufacturing, computer fabrication scheduling, fault diagnosis for telecommunications equipment, and production management for a wide variety of factories.

Modern solutions to CAD are in principle polyvalent, and as Figure 9-1 suggests, they include issues that range from product description to specifications and their equilibration properly defined, quality standards, coordination of process characteristics and fine grade integration between design and manufacturing engineering. This integrative approach calls for solutions enriched with artificial intelligence able to provide a significant amount of coordination.

We talk of solutions able to face the magnitude of the job that needs to be done. A basic principle in industrial and business life is that the choice of tools should be made to function for the specific problems being confronted. The problems grow because of:

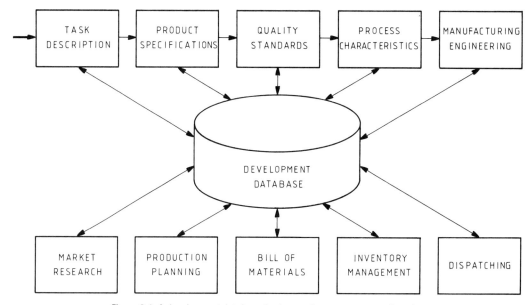

Figure 9-1. A development database for interactive concurrent engineering.

- The cutting edge of competition,
- An increasing sophistication in customer demand, and
- The amount of greater complexity that can be handled through technology.

Our ability to respond to them should match and surpass such growth. Otherwise, our company will not be positioned for survival.

Today, the solutions we provide should take a systems view of the environment in which we are working. They should not only answer our internal CAD needs but also the foremost requirement of interconnecting on-line with our business partners: clients and suppliers.

Networking with our business partners, our clients and our suppliers is the real reason to be in business. Sound technical solutions are needed and Figure 9-2 makes this point by presenting a three-layered structure:

1. A core CAD support center operation that includes optical disks, very high resolution input/output devices, the text and data warehouse, and supercomputer facilities for number crunching.

The reason for supercomputer support online to CAD workstations is that many design-oriented programs start being MIPS suckers. The 1990s will be characterized by complex artificial intelligence programs and software that manage digitized graphic images, and these use a lot more processing power than what is available through mainframes. Besides that, supercomputers provide quite cost/effective solutions.

2. A distributed environment of networked, AI-enriched CAD workstations, interconnected among themselves and to the core CAD support center as well as to all other information technology resources of the corporation.

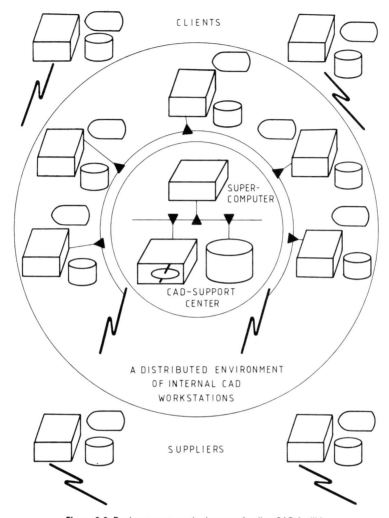

Figure 9-2. Business partners in the use of online CAD facilities.

CAD workstations that are not networked are not WS worth mentioning. No wonder that the WS networking business saw a growth of about 60 percent per year in 1988, 1989 and 1990. At the same time, systems integration requires legions of skilled analysts and software specialists to customize solutions assisting them through knowledge engineering in order to improve their productivity.

> 3. On-line realtime communications with the CAD systems of our business partners, supported as far as possible through common protocols and short of that by expert systems providing fully automated protocol translation, including graphics and data structures.

The second generation CAD rests on these three pillars and at times goes beyond them by way of interlinking with business partners. The product planning solution developed in Japan by Teijin Systems Technology presents a good example of this type of usage.

Starting with the fundamentals that characterize an engineering design process, the target *market segment* is first determined based on an analysis of customer age, occupation, and life-style. After a target segment is selected, a *product* for that segment is specified to a specially written AI-enriched system, leaving it to the latter to select a first approximation to optimal product specs for that market segment:

- Considering fashion trends,
- Including style themes,
- Reflecting on differences between the home market and foreign markets.

This example comes from the fashion and garments industry, but we should always learn from the best solutions wherever they are found transfering what we learn to other industries.

For instance, Mika Lady has written a personal wardrobe diagnosis expert system that uses the knowledge of fashion designers and house mannequins. This AI construct is used to design and/or select optimal outfits for customers considering their figure and fashion preferences.

Things come together, considering the developments in the fashion industry and in other consumer products involving electronics, electrical and mechanical engineering, in the domain of image processing and its capability to produce any combination of patterns. As the preceding chapters have outlined, pattern recognition can make an important contribution to any designer.

In the paper and pulp industry, for example, Toppan Printing has developed a corrugated cardboard design approach and Dai Nippon Printing has a gift assortment intelligent design system. Both have been put into practical use and both involve pattern-oriented approaches.

In another example, Suntory has developed a solution for the managers of liquor stores. Considering the location of a store, the expert system gives advice concerning the type of business best suited for that location and then it advises on store interior (a design issue), pricing, and service details.

In the electrical and mechanical industries, Sumitomo Electric, a nonferrous metal company, has aplied AI technology to fault diagnosis of local area networks (LAN) using optical fibers. In the general machinery industry, Daifuku has developed a fault-diagnosis system for machine tools while Toyota Machine Works uses AI technology both for computer-aided design and for production management.

3. APPROACHES TO THE USE OF AN INTELLIGENT CAD SYSTEM

For any product and for any process, the initial specification and the resulting criteria for a successful design are essentially responses to market needs and they reflect ever-changing values. Given an adequate specification and the means to modify it according to requirements, there is no reason why intelligent CAD tools should not be made, by area, for all areas of engineering design.

A far-sighted approach to fundamental design requirement is vital in terms of reaping competitive advantages as a result of investments in new products and processes. At the same time, market pressures see to it that value-added features are part and parcel of remaining a player, and this calls for steady refinements in implementation starting at the design level.

Four sorts of operations can be performed in an able manner through CAD: *Specification, abstraction, optimization,* and *testing.* All four can be enriched through artificial intelligence.

Specification involves a negotiation between alternatives as well as competing ends done by the designer and it involves a judgment about what constitutes a valid product. The designer has a concept but the CAD engine has to present specifications that are formal and verifiable. To work properly, this approach calls for a definition of:

- *Input:* CAD layout, part list, decision instructions.
- *Output:* Product and/or process plan.
- *Status:* Original design, production use, other state.

Input, output, and status have to be developed and specified in coordination with design issues such as robotics, planning the actions to take place in a detailed, well-documented manner.

The infrastructure for handling in an able manner this input + output-status process is shown in Figure 9-3. It rests on well-defined phases each with feedback to the preceding phase; it thus helps in fine grain refinements.

In the specifications level of engineering design, expert systems provide the ability to arrive at an interactive solution while trying a greater number of alternatives. They also make easier the duplication of expertise and its wider diffusion.

The use of artificial intelligence tools in this connection is promoted by the fact that knowledge engineering techniques are more readily accepted among the endusers because they show both knowhow and user friendliness. This is especially true of those designers who have perhaps viewed computers as "useful, but unable to solve real problems," the given reason typically being that design problems involve a lot of judgment.

Working within this context, an expert system would allow the designer to intervene in a controlled way with the design, interfacing between the enduser and the process of automatic generation. This depends on:

- The representation appropriate to the application, and
- The relative compatibility of this representation within the larger system.

The user might be able, in successive stages, to contribute his imaginative approaches to the developing design landscape. Knowledge engineering facilitates this intervention and also sees to it that as much as possible is made of past designs in the generation of new versions or even totally new products.

Mathematical models make an important contribution to the process of *abstraction* which is a fundamental activity in engineering design. We have known this for centuries and for over 50 years we have appreciated the assistance offered to the human mind through scale

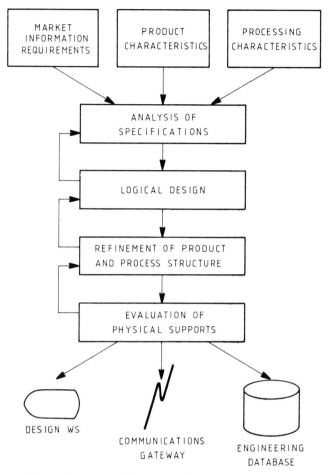

Figure 9-3. Analysis and logical design precede the physical design of products.

models, wind tunnels, and differential analyzers. Now we enrich CAD through digital simulators.

Simulation is a working analogy that can be of significant help to the processes of abstraction and design. As we have successively moved:

- From analog simulation
- To digital simulation,

we have enlarged the population of products and processes subjected to experimentation, as well as the size of the systems that we study and simulate.

But while in the past simulation basically concerned physical characteristics, AI tools now make it feasible to simulate logical processes. This is a new dimension in abstraction of which the design engineer must be aware to use it properly.

Essentially we are talking of an exciting field of research in a discipline that has come to be known as *cognitive engineering*. This is a hybrid of psychology and mathematics, where the theoretical and the arcane come to have practical applications.

Optimization, our third basic reference, is a more complex process than the two preceding processes are. Experimentation takes place to optimize a product, process, or system that was already found to be feasible.

As many companies have discovered through practice, expert systems can play a key role in optimization, and this exists for two good reasons:

- *First,* they can bring into play the way the real expert works in optimizing product characteristics, making his knowledge widely available to other engineers.
- *Second,* they help structure and formalize the methodology to be used, thus making a more homogeneous approach among designers feasible.

Logical methods supported through AI enable us both to employ and to control upward-propagating constraints. This gives the designer the opportunity to do with the expert system what is a cornerstone to design: feed back information. But it also provides him the added facility to study the difficulties in implementation details, particularly at higher levels of representation.

Testing is the final phase in every product. The evaluation of the optimized design has to be subjected to tests all the way from the product itself to the processes that will make it; that is the domain of manufacturing engineering. Here again, CAD and expert systems can help in assuring homogeneity and consistency.

One of the main aims in developing intelligent testing methods should be to isolate criteria for a good design. In doing so, reliability, the topmost issue, must first be brought into perspective with the AI construct paying due attention to it. Other criteria at the top of the list are able man-machine interfaces and the need for a supportive environment.

Through expert systems, CAD can be placed in a context in which the values used to produce judgments about what makes a good product are more evident. This enables the designer to:

- Identify salient problems, and
- Address himself to them in an effective manner.

In an era of increasingly complex engineering projects, there is every reason to believe that AI tools are necessary for technological progress. At the same time, their creation provides us with a rare opportunity to conduct real-life experiments with models of the complex cognitive processes involved in design.

Often it is not sufficiently appreciated how important knowledge-based considerations are in engineering design. Yet, even if we pay close attention to algorithmic processes and control structures, our project could fail in application if, for example, we do not assure able human interfaces. Careful consideration must also be given to the way in which a solution is introduced to its users and how it will affect their work patterns. As experience in the implementation of expert systems in computer-aided engineering steadily grows, we are able to address a growing range of issues.

4. SOLUTIONS TO BE OBTAINED THROUGH HIGH TECHNOLOGY

Sections 2 and 3 brought into perspective the fact that industry leaders are fast adopting information engineering solutions endowing them with knowledge-based tools. In doing so, they aim to assure that their designers:

- Effectively interact with the engineering database,
- Get easy access to text, data, graphics, images, and hence, the *multimedia,* and
- Obtain balanced answers to the queries they develop in the normal course of their work.

Figure 9-4 presents an integrative approach to *computer-aided design* detailing what comes under each of the three keywords. Designers who think they are best usually become the best if they are given steady training and the appropriate tools.

Both knowhow and tools must be steadily updated because what was high technology yestarday today becomes a standard feature. This is the use of a computer-enriched presentation with full screen editor capabilities and logical pathways to the solution sought by the engineering designer.

As we saw in the preceding section, among the advantages of an expert consultant system for computer-aided design are interactive explanation facilities provided by the expert system. These can be a significant aid to polishing an engineering design as they rest on knowledge elicitation from top professionals.

The able use of expert systems can also assist in closing the presently existing gap between the company's own design engineers and its purchasing staff. Since the purchasing executives rarely have an engineering background, they find it difficult to follow the diversity that now exists in sources of supply, but intelligent CAD can be of assistance, both in providing more accurate specifications and in flashing out:

- Differences in delivery schedules,
- Compatibility issues, and
- Quality assurance problems.

Expert systems can help create a homogeneous presentation environment, and they can also assist in implementing intelligent databases.* This is consistent with the fact that artificial intelligence focuses on developing programs able to perform tasks normally associated with intelligent human behavior that involves:

1. Engaging in dialog
2. Understanding natural language
3. Recognizing and synthesizing speech
4. Computer vision
5. Pattern recognition.

---

* See also D.N. Chorafas and H. Steinmann, *Supercomputers,* McGraw-Hill, New York, 1990.

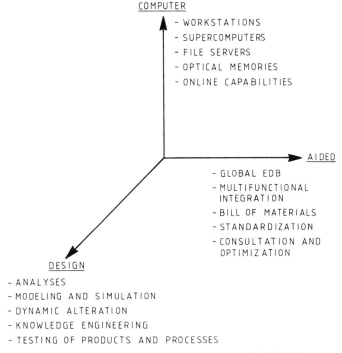

Figure 9-4. Computer-aided design is a multidimensional discipline.

A characteristic of high technology solutions is that through AI we manipulate symbols rather than numbers; make inferences and deductions from available information; and apply knowledge to improve design performance.

There are immediate possibilities for implementation closely linked to our developing knowhow. Evidence developed during the 1980s documents that the knowledge revolution impacts on engineering design—and therefore on our competitiveness—for four principal reasons:

1. The rate of growth in manufacturing industries created a need for better qualified engineers.
2. Fast-moving automation (including robotics) requires that skills are analyzed in a more fundamental way and are made explicit.
3. Product development cycles have been greatly shortened, thus calling for appropriate tools to handle compressed timetables.
4. Product diversification (rather than homogeneity) is the new characteristic of product development.

The market pushes along that way. Versioning a given product requires a greater degree of intelligence to see it through while a company remains a low-cost producer and distributor of products and services.

The Japanese auto industry has shown the way. Although between 1975 to 1985, the Japanese automakers focused on cost competitiveness through the use of technology (from

CAD/CAM to robotics), since the late 1980s they have adopted a new strategy: car *personalization*. This can be helped through expert systems usage as a layer higher up than CAD/CAM is. The key is to preserve the low cost of mass production while individualizing the product itself.

While the theoretical goal of AI is to make computers more useful by understanding the principles that make intelligence possible, there are practical goals that are concrete and attainable as well. For the first time, there is a consensus that knowledge engineering is a technology that can be applied to solve real problems:

- Today in isolated instances, and
- Very soon in a broad range of integrated applications.

An important component of continuing competitiveness is how we put into practical use our developing understanding of symbolic, nonalgorithmic reasoning processes. Also, important is how we represent symbolic knowledge for use in machine intelligence which is embedded in computer-aided design.

5 MANAGING OUR ANALYTIC AND SYNTHETIC SKILLS

The provision of advice through a consultant CAD program assists in the analysis of the design task; it also reveals the influence of logical processes. As such, it can be of assistance in the:

- Organization of the design team, in the
- Way the engineering mission is partitioned, and in the
- Coordination of manufacturing tasks.

Commendable results can be obtained by matching technical solutions to practical problems, leading to the consideration of fundamental issues such as design formalisms and representation.

Design has long been considered an ill-structured problem that is difficult to handle through software. But the foremost applications in CAD, as well as research into AI, have revealed the importance of using a knowledge-based logical representation of the problem as the key to making effective solutions a distinct possibility. This finds its counterpart in the way that good notation in mathematics makes thinking about quantitative concepts easier.

In other words, while over several decades we have at our disposal formal mathematical methods for automating design, with expert systems we aim to capture informal knowledge. Assisted through CAD software and AI, the designer is naturally led to the identification of areas about which not enough is known.

Many questions, however, still need to be answered: How can algorithmic and *heuristic**

---

* Heuristics is basically a trial and error approach. As such, it contrasts with algorithmic solutions which follow a predetermined path.

knowledge be combined? How do we elicit the knowledge needed for building AI aids for specification and for optimization? What kind of formal language should be used for internal representation? Representation has an important influence on logical problem-solving:

- The wrong representation can make even the most competent problem solver inefficient. (If you don't believe this, try doing multiplication with Roman numerals.)
- A good representation can help towards problem solving; it can also extend the power of a problem-solving strategy into new domains.

Through good representation, an expert system may be able to learn to do more than the original expert could achieve. The bottom line is that there is more than one way of acquiring knowledge from the expert. Methods of inducing rules from examples are being developed, and rules established by this approach have been found to be quite efficient in terms of performance.

Particularly in the fields of optimization and testing, *representation by example* can make previously obscure domain knowledge explicit, coherent and logical. This deepens and extends our grasp of the domain and explains the interest shown by engineering designers in using AI to help discover previously unknown algorithms for problems that can then be solved by conventional computing methods.

Both *analytic* and *synthetic* approaches should be used in this context. On the analytical side, for example, a user organization saw to it that for purposes of mass production:

- The range of skills possessed by a single craftsman were broken down into their component parts.
- These parts were then restructured into elemental actions that can be performed by robots.

As an example, Schlumberger's *Dipmeter Advisor* aids in borehole analysis in the search for oil deposits; helps in the interpretation of scant clues; projects information on the incline of a subsurface sedimentary deposit, and manipulates logs showing the degree and direction of the tilt (dip) of sedimentary beds.

Petroleum geologists studying the log will be looking for evidence of ancient reefs or channels in which hydrocarbon deposits could be trapped by impervious rock. The expert system transforms the measurements in these logs and adds local geological knowledge from its knowledgebank.

To automate reporting and other nongraphics database capabilities, the CAD workstation needs specialized applications software for creating intelligent vector designs with associated nongraphics data. Designers can become productive by using such output as a template for new products and processes:

- As changes occur, designers can overlay intelligent vector data on the raster image.
- They can then erase the obsolete references and gradually convert raster images to intelligent vector images.
- They can provide for themselves as well a smooth transition to the intelligent design environment.

The aim is to produce *answer products,* the Schlumberger phrase describing the information it sells to petroleum company geophysicists, geologists, and engineers. The Dipmeter Advisor incorporates the knowledge of the user into an interactive session divided into eleven phases; only four of these phases in the interpretation session use rules. The expert system:

- Codifies the expert's view of interpreting dipmeter data,
- Confronts the operator with a choice of conclusions in order of probability, or
- Displays all the possible conclusions to the expert user for his decision.

The user can add his own conclusions or modify the conclusions reached by the expert system. This is a combined effort with the AI construct interactively incorporating the knowledge of the user; it is also an example of how AI can assist in *synthesis*.

For the synthetic frame of reference, an expert system was built to coordinate body shapes, engine performance (and structure of an engine), the configuration of wheels, chassis, and so on. Such approaches have an evident impact on design methodology.

A similar statement can be made about the use of AI for better communication. For years it has become obvious to users of CAD tools that unless computer output was easy to understand and human input to computers was carefully constructed, the system did not work. Now AI helps provide a much more efficient man-machine interface by shaping a comprehensive computer-led dialog (the human window).

Sohio developed a seismic workstation for data preparation to help with the analysis of seismic data. An expert system by Battelle makes it possible to generate production plans for rubber diaphragms automatically. Alcoa is using an expert system that enables it to model and derive new alloy mixes more rapidly. Kodak is doing this through AI business planning and process control.

As these examples help document, design operations are modeled by functions that specify, generate, visualize, simulate, and refine. AI constructs help the designer to execute hypotheses, testing the tentative statements they involve:

- An engineering design is projected, modeled, tested.
- The design hypothesis is then modified according to the results of the testing.
- This takes place online assisted through simulators and heuristics.

As a result, the time formerly needed for engineering design is compressed significantly; better response to market requirements is assured; costs are kept under control; and quality is improved. Specification and refinement is done automatically at all-but-the-topmost level of representation, with better defined specifications being the result.

6. COMPONENT PARTS OF A VALID APPROACH TO EXPERT SYSTEMS

As the products and processes we handle through computer aided design become increasingly complex, they require intelligent planning and control strategies. The message given throughout at this chapter is that expert systems can contribute to both product design and systems solutions, provided we have:

- A model of the aggregate to be managed.
- A situation assessment function for interpreting input data,
- A clear understanding of functions representing desired activities, and
- Action-oriented ideas regarding the output.

A valid way of looking at artificial intelligence in CAD and its possible contributions is to look back in time and ascertain the benefits brought by CAD over what was known in the 1950s and 1960s as *scientific calculation*. The so-called scientific calculation was basically a manual design process, but as we were to discover, the manual design of some processes such as large scale integration (LSI) chips is practically impossible.

By contrast, through the able implementation of CAD technology, we were able to produce chips having tens of thousands of gates:

- But as design sophistication increased, a great interest became evident in improving existing CAD systems.
- At the same time, the more automated a process or product is, the better understood it has to be in order to lead to technical and commercial success.

System management, too, is an ideal area for artificial intelligence, as management functions require knowledge-based methods, including planning, designing, evaluating, and controlling activities. Expert systems help represent the elements of each one of these domains and their interrelationships, providing a basis for linking observable data to apparent system states through causal, diagnostic reasoning.

Such reasoning is not unlike that of human experience. When we observe certain data, we attempt to find the possible states of the system which the model suggests would produce those values. Figure 9-5 brings forward an example from network design; this implementation includes technical specifications, software supports, the bill of materials, reliability/availability, and cost evaluation.

Expert systems can support in an able manner future projection, examining one or more alternative situations to identify an objective outcome among options. The keyword is *synergy*:

- The expert system can project technical data more accurately and quickly than the designer can.
- The designer can often prescribe plan revisions that achieve desired results simply and efficiently.

Like CAD software, the expert system contributes some value-added characteristics from learning to reasoning and problem solving. This is a process that suits our knowledge-intense society well, as our output increasingly consists of information, services, and experiences:

- The domain where expert systems can be successfully implemented is mapped into a knowledge bank of facts and methods.
- Rules about a particular problem or application reflect what the human expert would do in a particular situation.
- A global data base that can be accessed on line enables of the engineering designer to do better work faster.

156 CAD, SCHEDULING, MANAGEMENT AND CIM

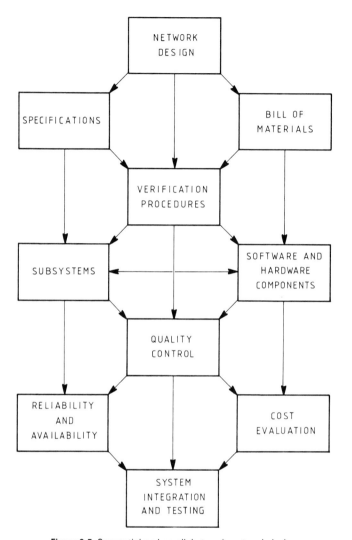

Figure 9-5. Sequential and parallel steps in network design.

Such methodology, however, implies a rigorous discipline. In terms of computers and communications, a knowledge-based system acquires a fourth dimension, beyond the three already provided by databasing, data communications, and data processing, of which the original CAD solutions have made ample use.

The introduction into CAD of AI technology has been helped by the fact that some of the most successful expert systems have been both PC based and interactive. The construct asks crucial questions of its users; it draws more tenuous inferences from its knowledge bank; and it reaches conclusions that emulate the way a professional would hedge his bets.

Unlike other types of software, the knowledge representation scheme in the expert system works with information flows and linkages that are transparent to the human user. At the same time, the latter is provided with a meaningful dialogue capability.

Such a dialogue between the designer and the product, process, or environment that he molds is of critical importance because an individual must approach complex situations in a comprehensive manner. Today:

- Thousands of manufacturers, for the first time, have to curb the toxic substances they spew into the air.
- To stop the destruction of lakes and forests by acid rain, utilities must halve the amount of sulfur dioxide and nitrogen oxides they belch into the air.
- Refineries have to slash cancer-causing fumes from gasoline.
- Automakers have to cut tail pipe emissions as well as make vehicles that can run on alternative fuels.

It is estimated that in the US alone it will cost about $25 billion to make the new clean air bill's mandates a reality. And the burden of paying that tab will land squarely on business.

The better the tools that are available, the faster and most cost effective will be the job that needs to be done. No matter how one wishes to look at it, AI-enriched CAD networked to risk databases is the soundest approach to executing the needed studies.

If we wish to appreciate the future trend in CAD in terms of expert systems, we should ask ourselves:

- What are the key developments in software and hardware stimulating AI usage?
- What are the implications for the user interface?
- How can information be represented in an expert form?
- What are the costs and benefits of using expert systems? What are the pitfalls?

We should as well appreciate the fact that, to be effective, a knowledge environment must be distributed. As a matter of principle, we should not operate centralized expert systems. This is particularly true in CAD. The artificial intelligence component must be integrated into every engineering workstation, and these workstations should be networked.

This poses lots of challenges not only in knowledge acquisition but also in supporting facilities such as inference mechanisms, database accesses, natural language dialogue interfaces, explanation facilities, and so on. One of the most vital references in an enlarged expert systems implementation in a design environment is the upkeep of domain knowledge, and this requires continuing support by the best design experts we have available.

7. PROFILE OF EXPERTISE IN ENGINEERING DESIGN

We said that expert systems require the collaboration of a domain expert not only during their *development* but also for their *sustenance,* that is, through their life cycle. What is the profile of an expert in the area which he masters?

- An *expert* is a person who possesses extensive knowledge in a finite domain.
- The logical process(es) this person follows in his thinking is (are) too complex to be described through an algorithm.
- Hence, we are interested in mapping it (them) into a logical structure that uses heuristics.

Done in an able manner, CAD expert systems help in the industrialization of design knowledge. One of their key features is the representation and processing of design knowledge regarding both products and processes in a specific domain. The fundamental difference between traditional data processing systems and expert systems lies precisely in the nature and treatment of knowledge:

- In DP, it is numeric, factual, and procedural, leading to computational solutions.
- With expert systems, it is symbolic, heuristic and declarative, providing causal relationships.

Because the best experts in our organization are going to formulate the rules governing their design skill, knowledge acquisition is very important: The expert systems' expertise rests on technical knowledge received from human specialists. This is particularly true of systematizing situation assessment methods and expanding the repertoire of design strategies and tools.

Designers of expert systems have found that dredging general rules from the subconscious of human experts is a huge challenge. Yet, it is a necessity. Expert systems must have a model of the domain of expertise; only then is deeper reasoning possible.

Knowledge and advice available in engineering design tasks from expert system consultation comes to the user through a question-and-answer session. During this session, the design manufacturing engineer describes his problem, which, without knowing it, often happens by way of *metaknowledge*. Metaknowledge is knowledge about knowledge.

Knowledge sources usually transform entries at one level of abstraction into entries at another level. Some knowledge sources operate bottom up. They aggregate several lower level entries into a smaller number of higher level entries. Other knowledge sources operate top down, exploding a compound knowledge source into its components. Metaknowledge handles knowledge about:

- Representation of objects (through schemata),
- Representation functions (function templates),
- Reasoning strategies (by means of metarules), and
- Inference rules (rules description).

Metaknowledge avoids the rigidity of always having to apply a hierarchical structure to an abstraction. It provides for flexible representation, and for this reason, it can make a significant contribution to expert systems development.

Concepts embedded in metaknowledge can be instrumental in knowledge acquisition as they are achieved by extracting and formalizing the knowledge of an expert designer for use by an AI construct. Examples of knowledge to be framed in rules are:

- Descriptions of products and processes,

- Identifications of relationships, and
- Explanations of procedural or heuristic rules

It is also important to underline the role of *symbolic computing* in engineering design. It allows for the processing of thoughts and concepts in addition to numbers, making possible the handling of knowledge- and reasoning-intensive activities.

"IF . . . THEN . . . ELSE" rules lend themselves to the representation of deductive knowledge—situation/action, premise/conclusion, antecedent/consequence, and cause/effect knowledge:

- The design rules are of the form: IF <evidence> THEN <conclusion>
- In automatic control, rules are of the form: IF <measurement> THEN <action>.

In working together with design experts, knowledge engineers should choose representation approaches that simplify the problem of encoding knowledge. They should also provide an intelligible encoding of that knowledge.

The challenge is one of making engineering design knowledge available to a larger user population, doing so in a transparent manner. Another aim should be that of constructing an expert CAD package that uses qualitative reasoning and symbolic analysis to give the engineer insight into the behavior of a design at an early stage.

10

Providing the Engineering Designer with Insight and Foresight

1. INTRODUCTION

One of the most challenging jobs in connection with the development and usage of intelligent CAD solutions is deciding the class of problems to be solved not only during the planning phase but through the whole life cycle of the operating system as well. A closely related challenge involves allowing for changes in configuration as design problems evolve with the passage of time.

In order to be flexible in terms of the solutions to be provided, we must estimate some of the more esoteric problems that can be encountered and must be handled by the intelligent CAD system and its users:

- We can effectively do so by projecting design applications and by using estimates based on pertinent past experience and realistic (but imaginative) assumptions and calculations about the future.
- Then, as CAD applications progress, we must recheck the consistency of our hypotheses and take the required corrective action, adjusting the facilities at our service.

Typically, such action should be aimed to assure that the engineering designer is supported all the way and that the tools at his disposal will help him become more insightful. He will be able to engage in in-depth analyses and will look further out in the horizon, anticipating change before it happens.

The magnifying lens in engineering design is an in-depth analysis supported by computational techniques for manipulating symbolic structures as opposed to only numerical values. In Chapter 9 we spoke of the contribution of AI in this domain and we also said that such symbolic computational approaches are becoming increasingly sophisticated.

Symbolic type design techniques are distinct from the standard computational solutions that give rise to numerical simulation. As such, they are used to create automated methods for manipulating the symbolic mathematical models upon which much engineering analysis and design are based.

The overall goal of the suggested work is to provide a foundation for creating computer-aided design tools that:

- Support the engineer throughout the entire design and analysis cycle, and
- Use expert systems to provide advanced symbolic reasoning capabilities.

The integration of reasoning methods with classical computation establishes the foundation for a very powerful approach to CAD, helping to solve basic design problems.

Such an effort will be so much more successful when the proper methodology is at hand, and when we have and use measures that can size up the effectiveness of the CAD system in terms of its attaining its various objectives. Many engineering studies can be so much more effectively conducted if:

- We have available for experimentation purposes a dynamic model of the product and/or process that uses a computer-based simulator, and then
- We will be able to run through this model a set of operational modes under estimated environmental conditions, so that
- We can enrich our experimentation through knowledge engineering, both in reaching conclusions and in justifying them.

Information learned from simulation should be fed back into the system design process to improve the stability margins as well as the reliability of the operation.

Gaining insight and foresight through AI-enriched experimentation, the designer will successively refine the results of error analysis. He will be using stochastic approaches for drives, boundary conditions and disturbances. He will adjust the design and its connections and parameters until the quality is satisfactory. This will eventually significantly improve his foresight.

2. UTILIZING AN ACQUIRED EXPERIENCE EFFECTIVELY

One of the best approaches ever devised for improving performance is to utilize effectively the experience gained at the forefront of computer technology in a wide range of design services, from the initial selection of CAD software to the full implementation of a total system. Correspondingly, the goal of *dynamic design* is to specify the structure and the component parameters so that the product or process under study will exhibit a specified behavior.

As the introduction has underlined, the specific issues of interest in any engineering design study include the stability, accuracy, response and reliability of a system. The basic procedure for determining sound design choices focuses on:

- Partitioning the product or process under study,
- Refining its specifications, and
- Assuring that quality and cost standards are fully observed.

The final design has to be validated by test methods. This process requires multiple views of the problem, ranging:

- From a high-level qualitative understanding of the behavior of the system,
- To a very detailed description of all of the components and their connectivity.

This process generally involves several basic steps: considering the behavior of each device; defining the component equations; describing how the components interact; and determining the state variables and generating the state equations.

The tools that can assist in this procedure are not only welcome *but* also absolutely necessary. As Table 10-1 suggests, some computational tools have been available for years, others are relatively new, and still others are currently under development. Among the latter are the heuristic approaches which become increasingly necessary.

Technology offers methods more cost effective than those we used in the past. An example is offered by General Electric who has a better alternative to the costly trial-and-error process of proving out a design with successive prototypes. Researchers at GE's R&D center in Schenectady, NY, have developed an expert system that tells engineers almost instantly whether a design can be manufactured efficiently.

Called *Sheets,* this AI construct can analyze a designer's work while it is still on the video screen of a computer aided design workstation. So far, this computer program has been used only to test the manufacturability of relatively simple shapes, but Chrysler Motors is joining GE's Major Appliance Business Group to help refine Sheets for more complex applications.

Along a similar line of reference, design engineers at Northrop Corporation's Aircraft Division have used AI to model the operation of a sheet metal parts manufacturing cell.

TABLE 10-1 Analysis Techniques and Computational Tools in an Engineering Design Process

Design Phase	Analytical Technique	Computational Tools
Making a rough sketch	Partitioning the problem and describing its key variables	Expert systems shell and graphics
Evaluating general behavior	Qualitative descriptions	AI constructs
Defining logical responses	Constructing symbolic equations	Expert systems shells
Generating state equations	Algorithmic and heuristic procedures	Bayesian, stochastic and possible models
Analyzing stability	Symbolic algebra	Customized programs
Simulating specific behavior	Numerical simulation	Simulation languages
Rapid development and testing	Prototypes	Expert system shells and other tools

Aspects of the cell's structure represented in the model include:

- *Concepts,* such as part baskets, machines, and part plans,
- *Concrete objects,* for instance, identification of a piece of equipment,
- *Attributes,* including capacity, contents, and destination, and
- *Relations,* including the upstream/downstream machine, the location of the part basket/contents of the machine, and so on.

Modeled functions include generating part baskets by work cell machines, moving part baskets from machine to machine according to plans, and machine failure identification.

A graphics editor was used in this application to create a graphic representation of the manufacturing cell. As the developer configured graphic images representing components of the work cell, the model editor simultaneously built an underlying knowledgebank of the system that is fully accessible to other facilities, including an AI construct for process planning.

This Northrop model is not alone in the realm of the manufacturing industry. Similar knowledge engineering solutions have been put to several different uses that include:

- Generating process planning approaches through expert systems,
- Testing the validity of alternative part production plans,
- Determining the optimal throughput of different shops,
- Assuring just-in-time inventories,
- Detecting bottlenecks in the production floor,
- Analyzing effects of different queues within a given scheduling scheme,
- Analyzing effects of machine failures on the overall schedule,

Invariably, the manufacturing industry finds out that the best production plan is the one which has been coordinated through the appropriate product/process synergy since the design stage. Increasingly, well-documented solutions use heuristics to supplement and sometimes substitute algorithmic approaches.

Used within the realm of computer-aided design, heuristics help the engineer to evaluate the natural stability of devices and systems, as well as to refine the problem on hand by specifying component parameter values. By contrast, simulation assists in creating a numerical model for experimentation and evaluation reasons.

Guided by basic engineering principles and past experience, the designer typically iterates basic conceptual steps to complete the specification of the physical system.

- In the early stages of the design, when the product or process has not been completely specified, the design engineer is working with a purely symbolic representation of the physical system.
- As experience accumulates, the problem-solving process becomes more concrete and more sophisticated solutions start offering sound, representative results.

Along this line of reasoning we can say that a dynamic model of CAD implementation (as well as of any complex business information system) has three key component parts as shown in Figure 10-1:

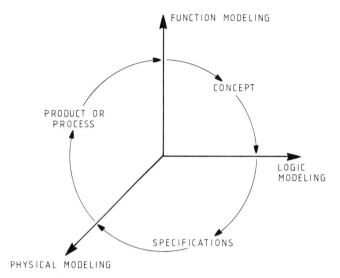

Figure 10-1. Modeling activities from original concept to realization.

1. *Function modeling* which identifies and describes participating endusers, functions, and their supporting communication paths as well as information stores.
2. *Logic modeling* which focuses on the detailed rules of behavior of the different functions and their use of databases and communication paths, plus the logical characteristics of the system.
3. *Physical modeling* which concerns itself with physical properties able to satisfy the logical requirements identified in step 2.

For instance, in the design of an information system, function modeling describes both manual and computer-based information solutions. One of its goals is to identify all the essential communication paths and the users of the information exchanged—from product specifications to the manufacturing process description and bill of materials.

At this step, another key component concerns all multimedia information storage (text, data, graphs, image and voice) essential to the product and the process. Activities still done manually have to be given equal rank with those that are computer-based; otherwise the study is skewed and partial. Indeed, the very reason for function modeling is to provide the conceptual framework by capturing business system descriptions in a systematic way, extending these descriptions into the elements of the logical model.

3. ENRICHING THE SIMULATOR THROUGH AI

Kayaba Industry, headquartered in Tokyo, is the leading supplier of hydraulic technologies. Its products include shock absorbers, hydraulic equipment, special-purpose vehicles, aircraft components, and marine equipment, with approximately 50 percent of the company's business being exported to more than 100 companies worldwide.

Kayaba uses AI and CAD quite actively with the process of designing hydraulic circuits divided into three stages:

- General circuit design,
- Static and dynamic analysis of the circuits, and
- Inspection of completed drawings.

These stages are normally entrusted to experts in each respective field. A dynamic analysis is typically handled by the research center staff, while the designing and inspection of drawings is handled by the designers at the factory.

As design engineers with expertise in hydraulic technology know through experience, one of the major problems with circuit design occurs when:

- Drawings are found which do not meet specifications, and thus,
- Much of the design and analysis must be redone.

In case engineering tasks falling into the above categories need to be carried out repeatedly, a great loss of time can then affect productivity levels.

In the case of Kabaya Industries as in that of thousands of other manufacturing firms, synchronization between the release of drawings and product specifications is even more vital when the company consistently develops new products for a widening range of applications. As a result, in analyzing the concentration of expert knowledge required for each task, Kayaba management decided to apply expert system technology to hydraulic circuit design operations, combining the processes of:

- Expert reasoning,
- Engineering design,
- Dynamic analysis, and
- Drawing inspection.

Each of these steps calls for a number of domain experts. The first AI construct Kayaba envisioned was one capable of capturing the combined expertise of such specialists, resulting in the best circuit drawings possible.

The polyvalent approach we have just described is mapped in Figure 10-2. For CAD implementation, this identifies the interaction between the simulator and the knowledge sources, the latter communicating with expert systems modules as many as the sources in reference do.

These knowledge-based modules compel specialists to apply their expertise to generate circuit designs as well as CAD drawings that meet well thought-out specifications:

- A simulation of a static and dynamic analysis is also carried out expressing the state of hydraulic machine components in the form of nonlinear equations that also
- Provide the basis for an analytic process consisting of two parts: a qualitative one modeling hydraulic circuits, and a quantitative one involving number crunching.

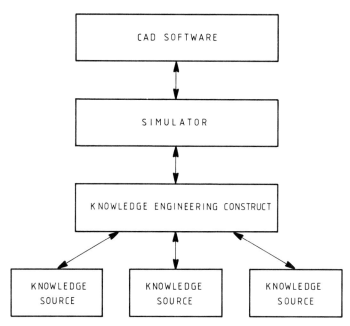

Figure 10-2. A polyvalent approach to intelligent CAD.

For logical consistency purposes, the model provides maintenance characteristics so that when hydraulic machine components are added or changed, constraint conditions can be easily expressed. Typically, such conditions affect results throughout the circuit system.

In a subsystem dedicated to oil hydraulic circuit design, the special circuit features are, for the most part, amenable to logical expression:

- Data relating to graphics images may be stored in the form of slot values, and
- Hypotheses based on deductive logic are generated and are regarded as viewpoints.

Values in the knowledgebank act as triggers for the instantiation and firing of rules. This is a critical point of difference between procedural and nonprocedural approaches. With the latter, when machine components are added or changed, values are automatically generated in the knowledgebank and logical inferencing begins on the basis of this new data.

Based on the drawing made of the hydraulic circuit diagram, an analytic model for the simulation of the circuit can be generated, duplicating the expert's qualitative experience in the modeling proper. Subsequently, a numerical simulation of the hydraulic circuitry is carried out, linked to the designing process and producing realtime simulation results.

Since there is a great variety of hydraulic machine components to choose from, selection is facilitated by a layered menu array. The composition of the circuit diagram is directly connected to the knowledgebank and this is true as well of the automating diagram inspection based on the specialist's experience.

Since the expert system's knowledgebank consists of encoded diagram inspection expertise, diagram inspection operation proceeds as if the needed human specialists were actually on hand to supervise the operation:

- The diagram inspection results and calculation of parameters are interactively displayed, acting as a guide for the designer.
- This makes so much more efficient the primary task of which we spoke at the beginning of this section: the *inspection of drawings*.

Classically, the drawing specialist checks to see that the functions of each circuit correspond to the specifications. This operation is deeply dependent on the special knowledge of the expert. Now the expert system sees to it that hydraulic circuits do not fail to satisfy the required specifications.

Kayaba Industry says that this solution has reduced the amount of work at the designer level by about 50 percent. If the work pertaining to analysis and diagram inspection is also accounted for, the result will save even more time. Furthermore, maintaining the quality of circuit diagram drawings at a specified level is now possible since the expertise involved in design and in analysis and diagram inspection has been regulated through the synergy of CAD and AI.

4. BENEFITS FROM THE USE OF EXPERT SYSTEMS IN PRODUCT DESIGN

The design phase of Vax 8600 demanded 4 calendar years of work at Digital Equipment Corp. The corresponding study of Microvax II and of Vax 8800 took 1.6 calendar years each. The difference was made up by the expert system which helped in the microprocessor design and by models that provided circuit simulation.

Due to increasing competition, a cut in the lead time is vital to profits and survival. The same is valid for optimization. Producing circuit boards can be time consuming:

- An industrial engineer needs about 20 hours to produce layout sheets that tell technicians how to assemble a particular circuit board.
- An expert system has been used to produce the sheets correctly in less than 90 seconds, with support costs drastically cut and changes rapidly implemented.

During the last few years, this process has also encompassed one of the key concepts: *concurrent* design and manufacturing. Concurrency, or simultaneous engineering, means that design and manufacturing engineers work together from the very beginning of a project to assure that it will not only do the job, but that it will also be cost-effective and of a high quality.

We are all familiar with elaborate designs that are either impossible to build, prohibitively expensive, or unreliable. Hence, simultaneous engineering can be used, accounting for *diagnostics* and maintenance early in the design cycle—as well as during operations.

An example from the implementation of AI-enriched online diagnostics comes from telecommunications. Since many different problems can interrupt phone service, to help locate the source and deliver a long-term solution, communication companies use expert systems for failure diagnostics. They establish the cause of the problem and avoid temporary work that causes additional, more costly problems in the long run.

The auto industry too has moved in the same direction. Derived directly from aviation, on-board diagnostic computers:

- Monitor technical problems,
- Remember the conditions under which the problems occurred, and
- Tell the automobile technician what has gone wrong, either in the shop or on the road.

The domain of services to be supported by onboard computers on motor vehicles is enlarging. Toyota offers an onboard mapping system in Japan. This video display shows a full-color landscape with the driver's car indicated as it cruises down the road. Volkswagen, Ford and General Motors are working on similar advanced navigation systems.

Nearly every car manufacturer is also investing in a collision-avoidance radar system, but at Nissan Motors there is already a production radar unit attached under the Maxima front bumper. Using knowledge engineering, it locates oncoming bumps and it adjusts the shock absorbers accordingly. Other projects focus on on-board front and rear radar devices that locate surrounding traffic or oncoming curves.

Ford Aerospace is working to apply artificial intelligence to automotive onboard electronics. Results indicate that a neural network of learning microprocessors could produce results way beyond those obtainable with today's comparatively simple closed-loop, realtime controls. Some analysts figure that:

- Ford Aerospace is working on 50 to 75 separate projects with the Ford and Lincoln-Mercury automobile divisions of its parent company, while
- General Motors counts about 150 separate projects currently underway that employ a technology transfer between GM Hughes, also a leader in aerospace.

This convergence of industrial sectors and their specialties is a reversal of past trends when, over the years, the aviation and automobile industries became increasingly specialized:

- Aviation, got involved with the comparatively limited production of high technology.
- The auto industry focused on the mass production of technology that is comparatively low.

During the 1980s, however, high technology became the common goal. Today's concerns about air pollution, energy use, global warming and transportation safety have forced automotive engineers and designers to shift into high-tech gear, sharing technology with the aerospace industry rather than adapting it years later or never. AI-enriched CAD has become a cornerstone of this shared activity.

The forces driving the high technology aerospace industry and the relatively low technology automobile industry have converged. Regulations, productivity, logistics, competition and worldwide operations are now parallel concerns common to both of them. This has resulted in a convergence of technologies with intelligent CAD at the hub.

As a result of such a convergence, we are now talking about a technology transfer in the design process, in the manufacturing process, and also in the equipment that helps us manufacture better products. Even satellite communications with dealers are affected.

It is axiomatic in the automobile industry that the difficult task is not one of developing new products, but rather one of adapting them to assembly line limitations as well as adjusting them to market drives. Intelligent CAD can help all the way, from conceptual activities to the production floor, and it has also proven to be an instrumental sales tool in the showroom.

5. A POSSIBLE STUDY OF DIESEL ENGINES

One of the best examples of a design engineer with insight and foresight in his work is a study jointly done by Japanese institutes of technology and Tokyo's Ship Research Institute. The subject of this study has been the control of marine diesel engines by means of the *possibility theory* and of the face patterns.

Japanese researchers have developed a new method to shorten the turnaround time required for optimizing control by using empirical rules some of which are expressed by fuzzy engineering.* The premise on which this research rests is that the fuel consumption rate (FCR, or simply F) of marine diesel engines depends on such factors as:

- The work load,
- The quality of the fuel oil,
- The fuel injection timing,
- The injection duration, and
- The inner pressure of the fuel pipe.

Control optimization of a diesel engine aims to keep FCR at a minimum regardless of the operating conditions; the gradient method, however, is the most popular for searching optimum points. Using fuzzy engineering, the Japanese researchers developed a new approach for estimating extreme points through sampled data. The following are the premises:

- A marine diesel engine's controlled variables are *speed* (N) and consumption rate (F).
- The manipulated variables are fuel flow rate (Q), fuel injection timing (T), and inner pressure of the fuel pipe (P).

---

* This section is based on my meeting with Professor Toshiro Terano, who is presently the director of the Laboratory of International Fuzzy Engineering in Yokohama. It reflects the results of the research conducted by Dr. Terano and Yojiro Murayama, Hiroya Tamaki and Fujio Inasaka of the Ship Research Institute, and Kenji Kurosu of the Kyushu Institute of Technology.

Conceptually, such a model is not new but as the design of marine engines improved, the machines themselves became much more complex, bringing to the foreground the requirement for automated monitoring.

To optimize control, FCR is measured periodically using as input the fuel flow rate and engine power. The gradient of FCR is obtained as a difference of FCR values at two different points of T. Then U is changed in proportion to the gradient:

- In the steady state, T and FCR correlate through a parabolic curve as shown in Figure 10-3.
- This curve might be suitable for gradient methods, if it were not that it is disturbed by noise.

Evolutional operation is one of the ways to apply gradient methods to a noisy process. This is essentially an interval estimation of mean value. However, if one reduces the amount of data in order to shorten the measuring time, the confidence interval is increased. As a result of this consideration, the researchers were led to the following solution:

- FCR is measured a few times at a fixed point of T.
- The gradient is obtained as the difference between the mean values of FCR corresponding to two different T's.

If the judgment of the gradient sign can be trusted, *then* the control process will depend on adjusting T in proportion to the measured gradient. *If* such a judgment lacks confidence, *then* the sample will need to be increased.

The applied second rule states: *If* the sign of the gradient is changed and the amount of data is more than three, *then* the FCR curve is identified as a parabola through regression

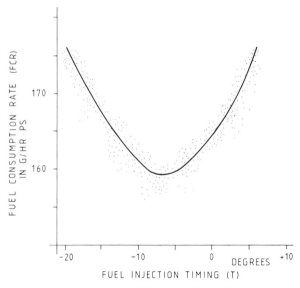

Figure 10-3. Fuel consumption rate versus fuel injection timing. A Japanese research project.

INSIGHT AND FORESIGHT IN ENGINEERING DESIGN

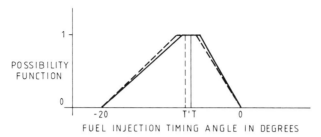

Figure 10-4. A possibilistic distribution for optimum point definition.

analysis. This permits us to find the optimum point of T, but the possibility of misjudgment is also increased and it can thus invalidate the optimizing point.

At this level, and in order to overcome such constraint, the researchers introduced *fuzzy engineering*,* Through it, they established a third rule: *If* the optimum point, estimated by the second rule is unreasonable from the viewpoint of the designer's common sense, *then* one should measure again and estimate the optimum point one more time. *If* the optimum point is reasonable, *then* move T to that point, and also modify the designer's knowledge considering this value.

The first and third rules just outlined are subjective and empirical; therefore, they are conveniently realized with a fuzzy controller. A fuzzy number is adopted to simulate the first rule:

- A measured value of FCR is represented with a fuzzy number of which the membership function is an isosceles triangle.
- The top of this triangle corresponds to the mean value of five values (measurements) and its base is equal to twice the standard deviation.
- The gradient is obtained as the difference between two fuzzy numbers by applying the extension principle.

When the membership function of the gradient is spread over zero, it is normal to judge its sign according to the area ratio (S)—where S is a *possibility function* addressing Rule No. 1.

The second rule rests on an ordinary regression analysis. To handle the third rule, the researchers introduced a fuzzy criterion that checks to see that the estimated optimum point is correct. As shown in Figure 10-4, the lower and upper limits of T are $-20x$ and $0°$ respectively. When T is over these limits, there are engine troubles.

The computation of the optimum point of T at half load is given with some allowance. Therefore, the membership function of the fuzzy criterion is represented as a trapezoid whose top center corresponds to the design value of T. *If* the value of this membership function corresponding to the estimated optimum point is larger than 0.9, then this point is considered reasonable and T is moved to this point. Such a criterion is revised as necessary.

---

* See also D.N. Chorafas, *Knowledge Engineering*, Van Nostrand Reinhold, New York, 1990.

Since it is necessary to transmit information effectively to operators on condition of marine engines obtained by instrumentation, the realization of a good man-machine interface is desirable as well. The face pattern method has been used in this study for transmitting engine conditions and characteristics, making it feasible to optimize monitoring.

For this purpose, the researchers provided an agile human window using face patterns. The face graph method helps in comprehending the characteristics of multidimensional information.

For instance, dynamic data on a marine reactor thermal simulator were expressed on a face graph in order to estimate its character. An experiment on the classification of expressions of face patterns was done to examine the relationship between the expressions of face patterns and the face variables themselves. This approach is just as applicable in monitoring marine engines as it is in monitoring whole plans.

Nine facial expression patterns shown in Figure 10-5 are typical of faces made by expressing two variables: the inclination of the eyes and the curvature of the mouth. Face variables can change over the range of possibilities; and the research by Dr. Terano and his coworkers included the dependability of face patterns in representing complex environments.

As these experiments document, the success of monitoring a plant's state by means of a face graph depends on how the plant's state variables are assigned to facial expressions and their variations:

- If plant variables cover enough information about the possible conditions, the facial expression changes according to the state of the plant.
- Since plant variables correlate with each other to reduce information about a plant's state, algorithms are necessary to represent facial expressions.

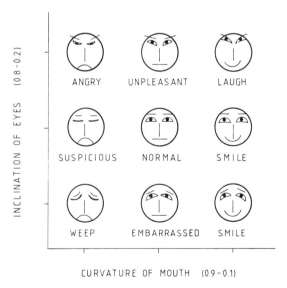

Figure 10-5. Facial expressions can be used successfully in a pattern graph.

Experimentation concerning information on an engine's or plant's status and condition monitoring of failures (which show dynamic symptoms) was done with a simulator. Here again, the condition of the failures was displayed on face graphs on realtime. Facial expressions changed as the condition of the plants changed with failures detected in the early stage, classified and properly represented.

6. APPROACHES TO IMAGE ANALYSIS

In Chapter 8 we looked into artificial vision from the viewpoint of robotics implementation at the factory floor. But it has also been said that we will return to this issue to consider its more theoretical aspects when we examine the impact of AI in engineering design to which Chapter 10 is dedicated.

The goal of *image analysis* is to recognize objects and relationships between objects. Objects are recognized by a set of features; the example presented at the end of Section 5 makes reference to the use of features for presentation and communication purposes. Overall, there has been an evolutionary path in image analysis:

- In the early 1980s, solutions were *model* based.
- By the late 1980s, they had become *rule* based.
- In the early 1990s, they are *knowledge* based, with knowledgebanks and memory-based reasoning (MBR) approaches.

Knowledgebank usage is fundamental to the development of adaptive systems able to tune themselves to their environment, sustaining a much greater accuracy than might otherwise be possible.

Figure 10-6 outlines a method of approach to the iconic phase of image analysis. This starts with *image acquisition,* for instance, through a TV camera, and continues with the databasing of the grey scale image. The next step is image enhancement followed by a feature-enriched grey scale that again is databased.

The feature vector is obtained through future extraction operations that may need to be preceded by segmentation and the development of a binary image. Key to this procedure is mapping from feature space to pixel space.

- There exists no unique approach for doing so,
- This is a highly heuristic field and the process can become very complex when there are many objects.

The only valid solution is the use of *a priori knowledge,* and hence of a knowledge engineering methodology. The more we have of this a priori knowledge, the more we can depend on what we expect to see. Then the task becomes one of *image interpretation.*

The key to an effective solution is enhancement of the features of interest. In this process, we typically try to separate the pictorial information of interest (that is, the signal) from the rest. This is a process of *segmentation* to which reference has been made in Figure 9-6.

Segmentation is followed by connectivity analysis. Here, too, there are no standard

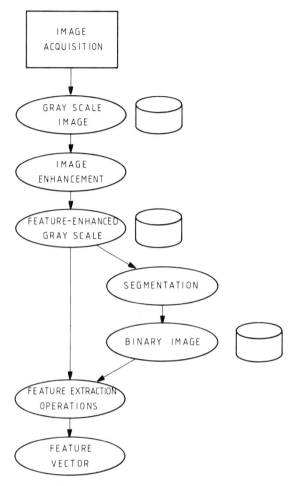

Figure 10-6. Iconic phase of image analysis.

procedures except the requirement that everything has to be done in high speed. Deadlines in realtime are dictated by the pace in which the manufacturing system operates.

Machine power positively relates to the resolution we wish to obtain. The more pixels an image has, the more operations we need, hence, the greater the speed required from the machine that we use. An equally important factor is the application perspective of the vision system. The following five examples focus on the capabilities that a vision system should provide for robots:

1. *Location,* for instance, of objects on a conveyor belt (for an industrial robot) or of a geological feature (for a cruise missile).
2. *Orientation,* so that the object may be grasped correctly.
3. *Identification* to find which of many objects is desired.
4. *Detail in Inspection* which is very important for quality control comparisons.
5. *Error correction* for manipulator positioning, feedback or navigation.

A factual and documented study should set goals as well as help sharpen the vision system's characteristics. When all features are properly defined, we reach the stage of *iconic analysis,* and from there the *symbolic processing* step. In other words, we can move towards a broader implementation of IA in design.

The emphasis on real time requirements should be kept well under perspective. The machine power required for iconic processing makes advisable the use of *supercomputers*. It also calls for heuristic and/or algorithmic description, the use of well-structured objects, and operations performed on small-time windows. Symbolic processing is typically:

- Non-numerical
- Declarative in a descriptive sense
- Involved in a heuristics search

Contrary to iconic processing which involves low flexibility, symbolic processing features high flexibility; it also requires a symbolic language (Prolog, Lisp) or expert system shell for its representation.

In most cases, iconic and symbolic approaches coexist, and to a certain extent they support one another. This has to be kept in mind when the feature extraction operations get structured and the design characteristics are set.

The best implementation of artificial vision systems in an applications-oriented sense is by *training*. We train the construct by showing it the work to be done, thus leading it to create its own knowledgebank.

As an example, we show the vision system a correctly printed circuit board (PCB) and ask it to store the proper connections. We then repeat this process teaching the vision system about the defects that may exist on the PC board. In practical applications environments, it takes eight to ten hours of PCB training to teach the expert system.

Such a method, however, has its limitations. In a specific implementation, for instance, such an approach worked well with an Apple computer that had one board, but it did not quite work with a mainframe computer that had 300 boards.

Hence, a better solution may be to feed the expert system with the CAD data base, reading the latter into the artificial vision construct. In an effort to obtain more valuable results, this constitutes a good example of the synergy that exists among CAD, the engineering database, AI and the process design.

An intelligent vision system must recognize and handle changes. A learning-based artificial vision methodology will suggest to the user the proper choices and solutions on all phases of an application—doing so in an interactive manner. Beyond this, the properly designed artificial vision system would dialog with its user.

Other design steps also have to be taken. For instance, in order to reduce false alarms while maintaining high failure identification, the preferred solution has moved from model to rule-based approaches, also introducing confidence levels and associated paths. Hence, inference engines have become an integral part of valid artificial vision solutions and of advanced design premises as well.

11

Expert Systems in Production Planning

1. INTRODUCTION

If the steam engine was the prime mover of the industrial age, the clock is central to postindustrial society. Computer-based planning processes are the information extension of the clock.

An orderly, punctual life actually made its first appearance in monasteries and many thinkers believe that minute clockwork is not natural to mankind. Planning is based on timing and as we will see in this chapter, it is a twentieth century invention. The First World has become so regimented by the clock so as to prove the aphorism: Time is money.

The paradox of time is that nobody seems to have enough of it, yet everyone has all the time that is available. All of us have 24 hours a day, but *time management* is a question not of how many hours and minutes we have but of how well we utilize them. No one can vary the number of minutes in an hour, but people who learn to live with the clock concentrate on planning their action, reducing the time lag between planning and execution.

Time is fast becoming the strategic factor in a manufacturing enterprise, and production planning is a good example of where the process of time management can be applied:

- The rapid development of automation and the prospect of electronically controlled production processes imply heavy costs for interruptions of any sort.
- One of the most important issues in efficiency is the time involved in planning, communicating, and executing.
- At a premium is the ability to identify a problem quickly and to deploy the organizational resources needed to handle it.

As we know it today, production planning grew out of the Gantt Charts designed in 1917 to plan war production. A similar reference can be made regarding the use of analytical logic and statistics, in other words, the use of quantification to convert experience and intuition into definitions, information, and diagnosis.

A cornerstone of production planning is *forecasting*—the ability to look into the future and to make an accurate prediction of oncoming demand. Forecasting is necessary because manufacturing lead times exceed sales cycle time.

Yet experience helps document that, in the manufacturing domain, our forecasts have been pretty mediocre, and this starts at the sales area. However, top companies have instituted forecasting systems starting at the sales level; their goal is the able *management of time* and the quantities associated with it.

Time is a unique resource. It cannot be accumulated like wealth, yet without it we can accomplish none of our tasks. Shakespeare said there are three things that do not return in life: An opportunity that has been lost, a word that has been spoken, and time that has passed by. We are forced to spend time whether we choose to or not, yet of all our resources it appears to be the least understood and the most mismanaged.

Through planning, we try to estimate the total effort for any given production project and its milestones in implementation. As we will see in this chapter, many approaches towards this goal have been adopted. The *plan by example* is one of them: The actual effort to be expended is charted out and the initial gathering of information focuses on quantitative factors representing the production effort in each area, including start-up time.

Once a feasible production plan has been established, accounting for products, labor, materials and machines, the job of optimizing begins. The word *optimizing* does not necessarily have the same meaning in every case, but it denotes a terrain where a number of rules can be established with certain assertion representing a finer or a coarser domain definition. We will return to this issue, particularly to those aspects that can be significantly assisted through knowledge engineering.

2. THE KNOWLEDGE-BASED MANUFACTURING ENTERPRISE

Production management is broad in scope including diverse activities ranging from personnel training to work sampling. Production planning is more limited in focus. It constitutes not only a plan but also its associated flow of intelligence within production management, providing a basis for scheduling as well as for control action.

The work of a production planning system is analogous to that of the human nervous system. Living organisms have the ability to move in directed ways and to accomplish intended results, but no coordinated movement of the parts of the body is possible unless the muscles receive nervous impulses. These impulses must fit a pattern; otherwise, aimless and useless movements will result.

The major duties of production planning are common to most manufacturing enterprises and can be described in the following terms:

1. Understanding the products and processes specific to the enterprise
2. Efficiently communicating with the sales department

178 CAD, SCHEDULING, MANAGEMENT AND CIM

3. Defining finished parts needed for sales and inventories
4. Determining material and components requirements
5. Timing all production, transport and storage facilities
6. Evaluating production schedules and costs
7. Determining machines and machine-loading requirements
8. Establishing effective operational sequences
9. Making up production orders and job tickets
10. Maintaining efficient inventories, preferably just-in-time ones
11. Helping to control quality
12. Assuring that all facilities for production will be available as needed
13. Coordinating transportation media and dispatching operations
14. Evaluating end performance

A production control system is efficient if it is the simplest one that assures the functioning of manufacturing facilities in the most economical fashion and without delays. But in an industrial organization the fabrication of products is not made for its own sake. Keeping a shop busy is by itself no indication that the profitability of the enterprise has been increased.

Production must be made to fit the needs of the market-place and this can only happen *if* a well-organized, realistic, far-reaching sales forecast can be provided. Planning, controlling, managing, and operating are activities linked to each other in a steady cycle as shown in Figure 11-1.

In the manufacturing industry, planning and controlling are multifunctional operations and they should should be approached in that fashion. Without the close cooperation between

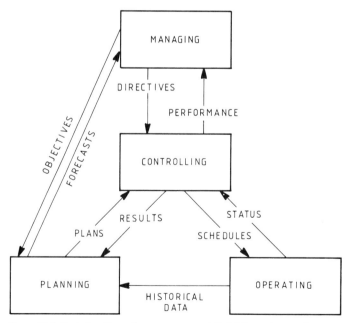

Figure 11-1. Basic functions of management and their information requirements.

sales and production, a company will never experience the smoothness of production, no matter how efficiently its manufacturing facilities may work.

Hence, the ultimate test of production planning is that it meets the needs of the marketplace with:

- Proper timing,
- At a minimum cost, and
- With high quality standards.

A realistic sales forecast requires algorithmic and heuristic models able to exploit marketing databases, as well as to handle inventories in warehouses and fabrication capacity at the production floor. Not only is the forecasting of sales a prerequisite to production planning, but also the correct analytical linkage must be developed to assure reasonably steady production schedules.

Production schedules are made in response to sales requirements, but they must also be closely coordinated with product design. This gives a glimpse of what is needed to achieve a *computer-integrated enterprise,* of which we will speak at the end of this book.

As Chapters 9 and 10 have underlined, design is a dynamic process. Planning is a sort of design with milestones as its kernel. Specifying a task, evaluating the constraints, establishing how it can be carried out, and elaborating on the proper timing are techniques of vital importance to the manufacturing industry.

From product design to sales forecasting, production planning, inventory management, and quality control, good business sense should prevail. Experience tells us that when business is expanding, stock tends to rise and extra capital is necessary to finance the increase. But we can do better than past experience by using knowledge engineering tools. That is how JIT has been put to practice.

Through production planning, management aims at utilizing existing resources (plant, labor, and finance) in the most profitable way. Therefore, it uses models and computers, introduces efficiency into the control cycle and provides for automatic feedback.

This, however, is only possible when:

1. There are clear policies regarding the use of productive resources,
2. Overall objectives take precedence over functional objectives,
3. Forecasting models are able to mark oncoming production requirements,
4. There are clear planning and control goals and policies,
5. Human resources, equipment and materials are on hand to substantiate the planning premises, and
6. There are a variety of tools available to assist and enhance the human intellect.

The use of expert systems in production planning is part and parcel of these tools that become that much more efficient if they are properly integrated as shown in Figure 11-2.

Algorithmic and heuristic solutions should assist management in meeting cost control criteria, helping to determine how to produce on schedule a given quantity of a specified quality, by using available production resources. Once the experimentation and optimization

CAD, SCHEDULING, MANAGEMENT AND CIM

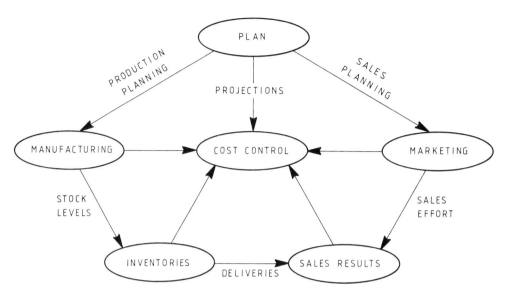

Figure 11-2. Cost control is at the heart of a business enterprise.

is done through computer-based models, networks must transmit the production plans online.

It is always advantageous if people and programs for ordering, producing, and selling interconnect online in realtime. This is true within the manufacturing company as well as between the company and its business partners (clients and suppliers) who both affect and are affected by production plans.

These are the guidelines that help to define the mechanics of the production planning system. The main component parts can be described in the following terms:

- *Scheduling* establishes the fine programmatic interfaces of the time plan, using available resources.
- *Routing* is basically the determination of the route for the movement of manufactured items through work posts in the factory.
- *Loading* includes the distribution of the required work load against the selected machines and workstations.
- *Dispatching* consists of issuing instructions authorizing the start of an operation on the shop floor as well as expediting goods to customers.

Replanning consists of reviewing routes, loads and schedules, in other words, rethinking when an operation is to be performed and when work is to be completed. For this reason, it is important that the system we adopt monitor planning and production processes with production planning information automatically audited but also distributed to departments according to their requirements.

There are common areas between cost accounting and production planning, two of them being purchasing and inventory control. In any business, common areas are in desperate need of good coordination. This is an organizational requirement. The computer will not improve through its mere presence old crumbling structures; it will only make them crumble faster.

A valid organization will see to it that full integration, from sales to inventory and production, comes into effect. In other words, the information system to be developed should stress the total integration of making mandatory data capturing at the point of origin.

Able solutions are always current, keeping pace with the progress of the physical processes being controlled and also providing a basis for experimentation. The line from *requirements* to *experimentation* identifies a modern, dynamic way of thinking which needs to be carefully sustained. In real life, however, this is rarely the case. Many present-day computer applications are still of the "discrete islands" type. Many companies incorrectly assume that a great deal of preparatory work has been done or will be done by somebody else and they tend to downplay organization. Invariably, this shows up to be most unwise.

3. PRODUCTION PLANNING THROUGH EXPERT SYSTEMS

We said that one of the promising domains to which artificial intelligence can be applied is factory planning, and the AI constructs run from simple to complex. For instance, it took a couple of days of steady work to create a simple expert system in manufacturing containing a dozen rules. The knowledge engineer:

- Started by listing all equipment this type of factory might possibly need,
- Entered this list into the program, as a protocol of alternatives, and he found that
- The options available for each alternative were at the center of the logic trails on the computer.

This tiny AI construct paid as a dividend a high multiple of its cost. To employ it, the user answers a series of questions with choices based on the type of products to be made. Each response moves the inference mechanism into a logical chain, leading to the selection of a machine tool. At any point in the process, the user can ask to see the rule upon which the program bases its answers.

Most real-life expert systems built for the manufacturing industry are in the general case more complex than that because it takes many rules to set up efficient production scheduling in a modern plant. Generating daily production schedules turns out to be a challenging proposition involving diverse and changing requirements and reflection on:

- Project priorities,
- Resource availability,
- Production quantities, and
- Product delivery dates.

Every one of these factors impacts on the generation of a daily schedule as well as on the longer term plan. The aim is not only to balance the production line but also to produce a bill of materials (BOM) and integrate it with the bookkeeping.

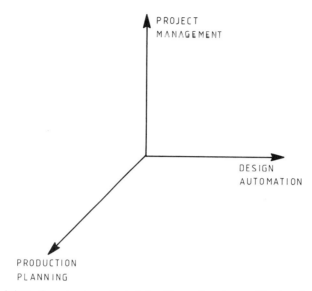

Figure 11-3. A three-dimensional coordinate in handling engineering specifications during hand-over to manufacturing.

It is precisely this greater sophistication of the solutions we are after which calls for an increasing number of expert systems. An equally important requirement is the provision of an integrative approach able to handle the passage of engineering specifications to production management. As Figure 11-3 demonstrates, this is a 3-coordinate problem: From project management to production planning, it involves the passage of product and process information among different departments: R+D, engineering design, manufacturing, production planning, and quality control.

The problem of optimizing manufacturing operations is more complex because manufacturing engineering and production planning theories change over time. *Line production* methods were established at the beginning of this century by Henry Ford, but already by the late 1920s and early 1930s, some American management pioneers such as Thomas Watson, Sr. at IBM, Robert E. Wood at Sears Roebuck, and Elton Mayo at the Harvard Business School began to question the way manufacturing was organized:

- The thesis of the challengers has been that the assembly line was a short-term compromise.
- Despite its productivity, it produced a poor economic state because of its inflexibility.

As a result, some people began thinking of and experimenting on alternative ways to organize the manufacturing process. This led to *quality circles* and the flexible computer-based production planning, representing the application of knowledge to the workplace—today we would say by means of expert systems.

Prior to applying technology to the planning and scheduling of manufacturing operations, we must acquire the appropriate tools. As a minimum we will require:

1. A domain- and problem-independent discrete event simulator that includes both analytic and heuristic problem-solving modules.

These modules will be expected to provide design advice, assist in selecting test cases, and interpret simulation results. The simulator should support alternate-scenario evaluation and facilitate verification.

2. An expert system shell for both interactive and automatic scheduling, as well as the proactive supervision of scheduled activities on the factory floor.

This AI construct should be analytic and heuristic—that is, it should use mathematical programming and knowledge representation tools—as well as constrained reasoning.

3. The means to assure a graphical view of the factory, allowing the user to specify goals and constraints and to make planning decisions.

Similarly, decisions made by the user must be evaluated with regard to how well they meet goals and satisfy constraints. Hence, the model should combine knowledge of goals and constraints with know-how about actual conditions on the factory floor. Only in this way can it help generate schedules satisfying the entire range of production requirements.

4. Supervisory routines to assist in monitoring deviations in schedules and advise on changes necessary to satisfy production goals and constraints.

If the granularity of the deviations is too large, the schedule will need to be replanned at a better level to match the prevailing constraints and requirements. Otherwise, realtime reporting could reflect the prevailing conditions.

5. A customized library of manufacturing domain knowledge to help in modeling flexible manufacturing systems, job shops, assembly lines, and so on.

Such a library should include expertise for analyzing capacity, inventory, idle times, delivery capabilities, and costs. Interfacing routines should also be available to help in interconnecting other domains when the need arises to redefine the parameters of computer-integrated manufacturing (CIM) approaches.

Some of the utility routines will have to provide expert level knowledge about the automation process itself. This will be of assistance in shaping, directing and facilitating communication among the subsystems in the factory environment to assure continuity throughout the manufacturing process.

Just as vital is the existence of routines for optimization reasons, along the path shown in Figure 11-4. Different approaches have been tested in this connection and they range:

- From linear programming,
- To scenario writing.

Assertions made in a production environment often resemble scenario writing which is one of the predominant approaches in knowledge representation.

A scenario writing approach is typically worked out with a minor number of rules, as the fine grain domain definition makes the larger number unnecessary. In fact, some of the rules might essentially be assertions in disguise. This is practically what is done with frames

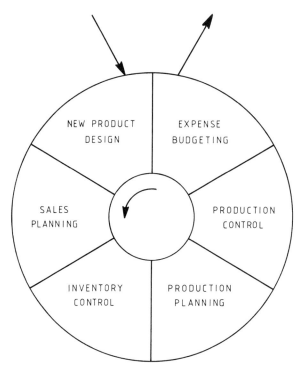

Figure 11-4. Steps which must be considered from new product design to expense budgeting.

(templates) whose interconnection can be interpreted as either a rule or a larger assertion construct.

4. AI IMPLEMENTATION EXAMPLES AT THE PRODUCTION FLOOR

Westinghouse Electric built a computer-generated planning program known as the Master Scheduling Unit (MSU). The original model was developed for the new automated manufacturing facility of a Westinghouse customer whose line of business is machining aluminum aircraft parts.

The object of this production planning expert system is to construct and monitor the execution of daily production schedules for the manufacturing shop. The Master Scheduling Unit will carry out distinct, but related, activities:

1. Generating production plans that reliably fulfill the requirements of the manufacturing plant.
2. Revising and maintaining the production schedules in response to shop conditions that constantly change.

3. Minimizing the number of late orders while maximizing the facility's machine utilization.
4. Handling priorites such as the assurance of on-time parts delivery.

As in all similar applications, knowledge from many difference shop areas influence the plan in reference, for instance, the type of hardware and software in the facility; the available human resources; the materials resources on hand; the machine capacities and types; as well as the degree of automation.

The shop floor in this particular implementation includes six numerically controlled machines, two-part inspection devices, pallet transporters, and a manned station for loading and unloading parts onto the pallets. There is as well a storage area for the pallets and tools.

The computer-controlled machines, parts inspection devices, and pallet transporters are all directed by networked microprocessors linked to computers running in tandem:

- File servers store machining and inspection instructions for the shop machines.
- The expert system resides on one of the computers and may be accessed by shop floor managers.
- Authorized production planners can change the priority of orders being processed.

The aircraft parts may require inspection in varying degrees on either side being machined, with the inspection requirements varying from piece to piece. Machining and inspection times can vary per side and piece.

With a lot-order configuration, the production plan must be able to handle many different machining and inspection times. This contributes to the complexity of scheduling, increasing the amount of knowledge that must be handled by the expert system.

Operating online, the MSU receives information on the status of any orders, as well as the status of all machines on the shop floor. Based on these references, it constructs a schedule of machining and inspection operations for the next three shifts. As these shifts take place,

- The AI construct monitors the success of the operations by querying various machines and the load station for current status.
- When necessary, it reschedules to change the priority setting of a shop order or to improve operating performance.

Priority status is assigned using a nine-position rating system that facilitates arbitration in conflict situations. Whenever possible, MSU postpones lower priority machine operations to allow the priority order access to required resources.

Machine failure, defined as the inability of a production facility to respond to the MSU's query, is one of the factors that will cause the MSU to reschedule. As in every factory, the planning process is not independent of the other activities on the floor, but it is highly dependent upon information from all areas and the decisions being made there.

The MSU expert system relies upon fast assimilation of data to make use of optimization algorithms regarding manufacturing resources. The same is true of ILOG, another production planning AI construct that helps to manage:

- Materials availability,
- Plant resources,
- Transportation costs, and
- Configuration know-how.

ILOG particularly focuses on configuration management. It does so through some 1,000 rules embedded in the complete model.

The equilibration of production lines is not the only domain where scheduling expert systems play a role. Hitachi developed an AI construct for work scheduling in supermarkets. It has been in operation since 1988 and it addresses itself to part-time workers, working in tandem with another knowledge engineering module in merchandizing. The latter's goal is to identify product sales popularity and help in the management of inventories.

This fairly sophisticated sales-line-balancing expert system provides an appropriate level of presentation and definition in accordance with its user's level of sophistication, specifically, whether he or she is:

- A novice,
- Somewhat experienced, or
- Well trained.

This reference to the expert system's self-adjusting capabilities is valid in terms of the commands made available to the man-machine interface, depth of help message, support of icons, and prompts.

An evaluation of obtained results indicates that the expert system in reference has met its stated objectives. The same is true of ILOG and MSU in terms of generating efficient schedules that can dynamically adjust to changes on the production floor. All three examples assure an online response.

5. VAGUENESS AND UNCERTAINTY IN PRODUCTION PLANNING

When in the early 1950s Dr. George Danzig advanced the theory of Linear Programming (LP), production planners embraced it as one of their most powerful tools. While LP continues to be a favored means for planning purposes, many manufacturing organizations now look to heuristics to supplement the algorithmic approaches linear programming provides.

In Section 3 we have seen a few examples along this line of reasoning. The mission of other expert system applications in a production environment is to:

- Generate, revise, and maintain consistent production schedules in response to shop conditions that constantly change.
- Assist in managing change, adjusting an active schedule to accommodate new events on the shop floor as well as the orders' arrival.

- Produce work instructions and plans for the production of parts and their quality inspection.
- Produce work instructions for the assembly of components, subassemblies and assemblies.
- Facilitate utilization analysis and resource planning that identifies conflicts and overloads during a manufacturing process.
- Plan and execute production routing in connection with a robotized production line.
- Monitor fluctuations in the plant system to predict malfunctions and identify problems as they occur.

The range of these functions tends to suggest that a great variety of applications in the production floor are handled through AI, and this is true. What such references, however, don't say is that there is a growing amount of *uncertain reasoning* incorporated into production planning modules. This has become necessary because of issues coming up ad hoc which can only be approached through the application of fuzzy engineering.

As competition intensifies, product innovation accelerates, and leadership in manufacturing is counted in terms of flexibility; an aggregate of planning strategies is needed to face the production planning problems as they develop. Among the factors impacting on the dynamic nature of this domain we notice:

- Changes in the size and in the culture of the fabrication work,
- An increased use of robotics and other automation tools,
- A need for flexible production schedules,
- The growing interest in swamping accumulated inventories,
- Emphasis on quality assurance,
- The development of supplementary product lines, and
- Overriding cost control considerations.

All these factors might have been present in years past, but their synergy was not as pronounced as it is today; moreover, the stated factors did not vary individually with the speed they do today.

As a result, parametric production planning and the development of experimental models are coming into focus. The same is true of heuristic approaches with the goal of:

- Estimating expected demand,
- Looking at realistic cost structures,
- Accounting for a number of planning variables and constraints, and
- Reducing the hiring, training, and firing of employees.

Fuzzy engineering is being tried in production planning because it has given commendable results in other areas involving process control perspectives. The goal is to face vagueness and uncertainty in sales schedules, avoiding the need for a great deal of rescheduling at a plant site—which is costly—as well as algorithmic computational time which may be long.

For instance, one of the applications of fuzzy engineering in manufacturing involves the search for production levels associated with:

- Sales projections,
- Inventory levels,
- Machine utilization, and
- Personnel costs.

Possibility theory, that is mathematics that makes use of imprecision, fits well within a flexible manufacturing environment. This is commensurate with that fact that models are only as good as their designer's perception is. In principle, in many production situations we can talk about improved results as opposed to optimal results. In a production planning situation involving uncertainty the:

- *Internal values* will be fuzzy, with the model reflecting a smattering of feasible solutions.
- An envelope of *upper* and *lower bounds* will act like confidence intervals.
- The answers will *not be crisp,* but fuzzy answers can provide a good approximation to real-world situations.

A production planning model has been done along this line of reasoning for a process type manufacturing operation. Its primary objective has been to smooth production line loading of noninterchangeable processes that produce components or subassemblies common to an intermediate semifinished goods inventory.

The first module in this expert system starts with a market requirement forecasting and does a Bill of Materials explosion from the product material standards contained in the data base. This produces an initial production schedule that may not be unsatisfactory if production levels vary widely thereby leading to a suboptimal utilization of labor and equipment.

The precision of the production plan was not improved, but its accuracy was significantly bettered. By using fuzzy engineering, week-to-week changes were minimized and so were the changes in the in-process inventory.

This solution was further enhanced to make a cost trade-off between labor, equipment and inventories, linking them in a materials requirements planning system. The inference rested on three pillars:

1. *Representation* of points (which are *crisp*) and interpretation intervals (which are *fuzzy*).

Such representation typically includes both factual knowledge and expert knowledge. Both were used in this model.

2. *Selection of a rule* which involves a search procedure as well as linguistic approximation.

Linguistic approximation is very important in this context. In a properly designed model it will typically express concepts that are in majority fuzzy.

3. *Inference proper* through pattern matching and analogical reasoning.

A problem area this project ran into is knowledge acquisition, particularly in face of its inexperience with novel situations and the inability of the production planning specialist to express his or her rules. "The more expert a man, the more difficulty he finds in expressing his knowledge," said the knowledge engineer in charge of the project.

6. SALIENT PROBLEMS IN PRODUCTION CONTROL

Thirty years ago, industry was just beginning to consider the possibilities of automatic production techniques, and planning solutions were commensurate with that early state of the art. Today, computer-based production planning schemes provide us with valid answers to the question of how to obtain a sufficient quantity with a satisfactory quality of the given resources, but they also require a number of prerequisites.

Aside from reducing costs by accelerating production rates, automating the production line and aiming at a greater uniformity of products, we have also to look into flexible manufacturing schedules as well as improved and simpler quality control procedures. Our products must be characterized by an improved reliability, reduced man-power requirements and production around the clock. To reach such goals, the three areas of production:

- Product design,
- Processed materials, and
- Fabrication machinery used

must be considered together as a system when robotics and any other type of automated production is contemplated. As we have seen in Chapters 8 through 10, each of these areas must be properly evaluated while we consider the multitude of related factors governing compatibility of the individual operations in the production line.

The following basic requirements have been established for a flexible production system to meet modern economic and technical requirements, and in each step knowledge engineering can play a determinant role:

- The automated line must be suitable for both large and small production lots, easily adaptable to design changes, and capable of building reliable products.
- The line must be flexible in handling a wide range of work and materials, so that it can be rapidly converted from the manufacture of one product to another.
- Production planning must have sufficient lead time so that new orders as well as materials and components may be used practically in an ad hoc way.
- Every measure should be taken to insure the high availability of the automated production line and its robots.

As it cannot be repeated too often, the effective use of automated production techniques requires much closer working relations within a given plant, between the market analyst, the researcher, the design engineer and the production specialist. The same argument is true regarding associated functions of the production control chores, such as inventory management and dispatching.

The National Dispatch Router (NDR), for example, is an artificial intelligence construct designed to support the role of the central dispatchers at Digital Equipment's Distribution Center in Northboro, MA. This expert system can:

- Devise suggested routes and schedules for freight transportation, and
- Minimize distribution costs, highway time and carrier contracts.

As in many other corporations, DEC's central dispatchers have to make quick transportation routing decisions based on increasingly complex information while they are looking for the most cost-effective way to get shipments transported. Routing optimization can be done neither manually nor through classical DP.

The dispatchers must keep track of the number of trucks in transit, the routes these trucks are taking, the shipments they are carrying, and the deadlines they are trying to meet. At the same time, they must balance the terms and conditions of the carriers' contracts.

As business grows, so does the volume of the dispatchers' tracking and decision tasks, this position requiring both experience and expertise. This fact has been at the origin of DEC's search for ways to:

- Capture the dispatchers' knowledge on line, and
- Automate the system in order to cope with the ever-increasing volume of transportation business.

That is where AI came in. The expert system approach proved to be the ideal way to capture the expert dispatchers' knowledge, providing the other dispatchers with a very capable online assistant. Designed specifically for DEC's distribution center, NDR assists professionals in developing quicker, more effective transportation schedules. In addition to cost benefits, the National Dispatch Router also:

- Brings to the dispatching process a consistency of performance.
- Increases the quality and ease of information flow—from decision to execution.

Not only does the expert system provide a ready backup for the current dispatchers, but it also assists in the training of new dispatchers. And it can easily handle the ever-increasing volume of distribution business.

Other manufacturing companies have done similar types of AI breakthroughs focusing just as much emphasis on inventory control. In Part One we spoke about the use of AI in attaining just-in-time inventories. Some other examples can explain this further.

A leading manufacturer has written an expert system that assists inventory managers in purchasing supplies by reviewing the inventory status and the:

- Current, forecasted and historical demand,
- Shelf life of each product,
- Location of suppliers and consumers, and
- Storage/shipping charges.

Through experimentation and optimization, the AI construct recommends modifications to the requisition and gives other advice on bettering inventory management.

Through state-of-the-art inventory systems, including realtime communication with suppliers, General Motors says that it can reduce its stocks by at least $2 billion, cutting pretax costs by $200 million annually. This is one of the reasons why EDS implemented a worldwide private communications network for GM.

The message to be retained is that expert systems in inventory management have found a wide area of applicability going well beyond the production floor. NASA has built a *software inventory and reuse* expert system that maintains NASA's complex applications inventory and gives advice on program reutilization.

Typically, these are model-based approaches that operate interactively, compare simulated results to current performance, and take maximum advantage of available knowledge about running bills of materials. To map that knowledge into the system, they incorporate heuristics for management reasons but also for failure analysis, acting in an "assistant to" function that increasingly gives most commendable results.

7. USING KNOWLEDGE ENGINEERING FOR PROCESS AND LABOR MANAGEMENT

The Ford Motor Company has built a multiphase Manufacturing Process Planning System (MPPS) to assist its production and planning personnel. Management considers this solution as having profound implications for the process planning activities at the production floor.

The solution is designed to provide the foundation for several knowledge engineering constructs aimed at improving the assembly process. One of its major objectives is to:

- Achieve standardization in *process description,* and
- Improve the clarity of *process sheets*.

At Ford, the process sheet is a critical document, the primary vehicle for conveying assembly information from the initial planning stage to the assembly at the plant level. As such, it is an effective means of communicating information at all stages of manufacturing.

Another goal is to support the creation of *work allocation* sheets by automatically generating detailed assembly instructions. Work allocation contains instructions assigned to the individual assembly worker.

An element analysis subsystem has been incorporated with the objective of generating a set of atomic work elements representing the direct labor content implied by the sheet. This is essentially a process script for performing assembly tasks, defining direct labor for all the actions that contribute to the product assembly process.

A third objective is to automate the process of providing consistent and accurate estimates of direct and indirect *labor time* involved in the assembly process. A nonproduct time is basically the indirect effort needed to carry out the direct operation. The automation of clerical and routine tasks frees the industrial and manufacturing engineer to use his expertise effectively in analyzing more intricate processes.*

---

* See also Chapter 6 on CASES by IBM.

The next goal involves the identification of significant contextual data: the vehicle and functional subsystem of the vehicle, as well as the establishment of plausible defaults. The next phase is to map individual statements on the sheet into the corresponding operations concept within a given taxonomy.

With the necessary variable bindings established for the script, its constituent activities are in turn classified and their standard counterparts identified:

- This process continues until a set of terminal scripts has been identified, and
- The result represents the direct labor content of the assembly information needed to undertake the operation.
- The expert system takes care of all details.

The fifth objective is to assure the foundation for *metalevel processing*. Process sheets are written at a fine level of granularity. But generic approaches can provide process descriptions at a higher level, thus promoting standardization in the assembly line across vehicle types and making it feasible to develop process sheets at a significant level of abstraction.

A further aim is that of setting a foundation for machine translation to reflect the fact that car production is increasingly an international effort and the overhead involved in translating process instructions into different languages can be considerable. This knowledge engineering project promotes a standard language designed to facilitate translation. The taxonomy is structured in a way to serve as an interlingual form of manufacturing communications.

As this last reference helps document, crucial to the whole AI project has been the development of a process description language. This is the required proper identification of a standard format for writing *process descriptions* which had been previously recorded in free-form English.

In the reengineered system, process sheets are described using a *case grammar* developed to meet the requirements of car assembly. Such language is essentially a dedicated shell that assures the expression of imperative English assembly instructions accessible to thousands of engineers with different educational levels.

The solution is taxonomical and this is consistent with the fact that at the kernel is a taxonomy of automotive assembly expertise as shown in Figure 11-5. This object-oriented approach maintains descriptions of concepts and requires an engineer to:

- Interpret the surface language, and
- Generate the implied atomic work instructions.

Concepts in the taxonomy include parts, processes, equipment, standard operations and geometric workstation models. The classifier is incremental in nature.

The overall solution is flexible. As concepts are added or modified, only those inferences that are required to position the current issues and to reclassify any related others are made. Two knowledge engineering interfaces are provided:

- A procedural one that supports reasoning within the process sheet and an end-user knowledgebank update.
- An interactive graphical inference that helps in taxonomy maintenance for system development.

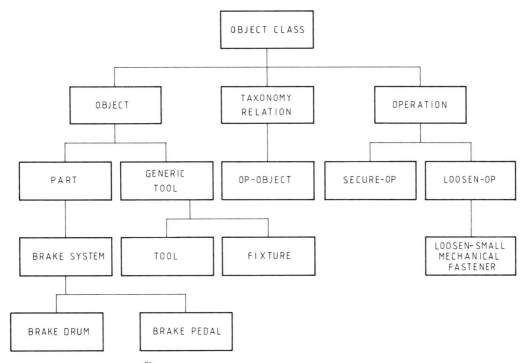

Figure 11-5. Object classes and taxonomical considerations.

The concepts, objects and their level of detail are determined by the requirements of any operation and the need to discriminate between operations. This also determines the granularity of the resulting plan and hence the resolution of any associated time estimates.

There is as well an assembly process simulator that elaborates the set of direct elements generating any additional indirect work objects required to implement the production plan. The first stage in this process is to identify the standard workstation configuration in which the assembly operation is assumed to take place from a taxonomy of standard configurations. This is done by:

- Creating a workstation description from contextual information in the sheet and classifying it, and
- Seeing to it that each work element is processed in sequence, identifying all milestones.

In carrying out the needed assembly, the operation is assumed to relocate itself between milestones. Such milestones are identified by creating concept descriptions and classifying them, thus permitting the handling of procedural issues and their corresponding level of detail.

12

An Infrastructure for Production Scheduling

1. INTRODUCTION

A knowledge-based approach makes it feasible to deal with a far more extensive manufacturing know-how while, at the same time, providing the ground for sound planning solutions able to respond quickly to the changes that constantly affect production schedules. These are the reasons why the foremost companies adopt the proactive approaches we saw in Chapter 11.

In all but the simplest manufacturing situations, the information base has become much too large and too dynamic in nature to be handled by conventional, data processing type scheduling systems. These by now traditional approaches produce schedules that are neither accurate nor able to respond to changing conditions at the factory floor as a result of a steady flow of sales inputs.

Competitiveness requires a superior *methodology,* and this statement is just as valid regarding the need for *organization* and *coordination.* We are not going to solve the manufacturing problems by throwing money or technology at them.

- Organization is the first basic ingredient for an effective solution.

In order to approach this issue in an able manner, we have to establish beyond doubt what an efficient production planning and control system requires—not in an abstract manner but within *our* operating environment which definitely includes the market's wishes and drives:

- Coordination is the second basic ingredient and it implies prerequisites.

To handle manufacturing coordination in an able manner, we need a logical model encoding the topological and methodological description of the production facilities. It should be built around data points and interact with the database. The manufacturers should be able to experiment with an AI construct reflecting objects underpinning the concept of interdependent production zones.

Figure 12-1 outlines the key modules of a scheduling system that benefits from knowledge engineering support. Designed to operate at the production floor, the construct can be easily extended to cover incoming marketing data, thus making the adopted solution flexible and sensitive to sales results.

Chapter 11 underlined that not only the forecasting of sales is a prerequisite to production planning but also the correct analytical linkage must be developed all the way from customer orders to fabrication in order to assure reasonably steady production schedules. Business variations affect inventories and scheduling in both ways:

- When business is expanding, stock tends to rise correspondingly. Extra capital will thus be needed to finance the increase; at the same time a burden is placed on the factory to revise schedules.
- When the volume of business contracts, capital from stocks is released only after a long delay. As sales fall, the stock movement slows down and so does the production plan.

Unless such changes are anticipated, there is a delay before the rate of in-house manufacturing and of suppliers' deliveries can be adjusted to the new level of requirements. If the reduction in business activity is general, suppliers' delivery times tend to shorten and outstanding orders are filled more quickly than is expected. In this case, too, capital

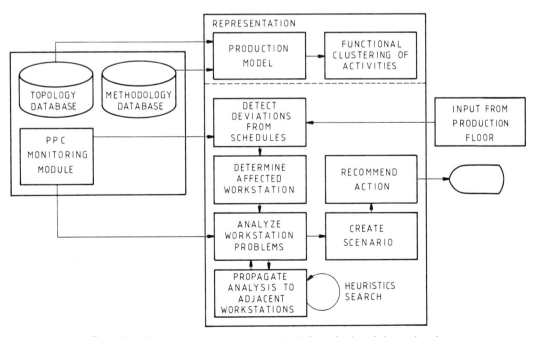

Figure 12-1. A production planning and control solution using knowledge engineering.

requirements increase negatively affecting the balance sheet. What can be done to remedy these situations?

2. CONSTRAINTS IN STABILIZING A PRODUCTION SCHEDULE

Scheduling is a complex constraint-satisfaction problem that can be effectively assisted through technology. However, if real scheduling expertise is not available in the firm the expert systems to be developed will be suboptimal—and the same is true of sales, inventory, and production coordination.

A sales, inventory, production coordination is very important because when other sectors of the business get optimized, as for instance, through just-in-time inventories, past solutions in the scheduling domain fail. Much more powerful planning approaches are needed.

Computer-based mathematical models can help in keeping production schedules flexible at the same time as they reduce changeovers. They can also be instrumental in providing sustained reductions in inventory while the overall customer service is improved and the manufacturing plan stabilized. Stabilization means:

1. Properly balancing the production schedule
2. Doing away with very frequent setups
3. Minimizing average delays per order
4. Significantly reducing work in process per order
5. Driving towards zero inventories
6. Largely improving the outgoing quality level.

These six goals may sound as good as apple pie, but there are an equal number of critical factors working against them: Complexity, uncertainty, suboptimal organization, a lack of coordination, incomplete information, divided production structure, and a horde of other constraints.

Complexity manifests itself through the number of different products that we make, their corresponding manufacturing processes, and the variety of personnel skills available. Solutions to the problem of complexity promote the need to develop alternative production plans; they also make it mandatory to proceed through experimentation prior to resource allocation, particularly in the face of contention.

Uncertainty involves issues such as resource availability and unavailability. This may be due to machine breakdown, retooling, tardy materials or other reasons:

- Some events may be external such as the oncoming job mix that results from customer orders.
- Others relate to internal scheduling, for instance, lead time and the need to accommodate urgent production requests.

Uncertainty is present as well in terms of process performance, whether this is due to robots and other machines, to operators, tool wear, materials, or failures in the schedule itself.

Uncertainty is an important factor in line balancing inasmuch as it impacts on the precision of *predictive* scheduling: It limits both the horizon and the temporal detail. It also leads to reactive scheduling thereby affecting the number of reschedules or even invalidating the scheduling algorithm altogether.

Suboptimal organization sees to it that the plant's ability to respond to uncertainty is diminished. This situation is frequent enough because most organizational structures have been set up on the wrong premise that there is certainty in terms of:

- Incoming information, such as sales orders to be translated into production orders, and
- The ways and means of dealing with the variety that inherently exists in market responses.

There is nothing more untrue than this sense of certainty, and the more we add layers in the organization to face the resulting challenges, the less we are able to respond to unexpected situations.

The impact of uncertainty and vagueness on organization and structure has attracted very little attention so far. Yet there are ways of dealing with them through *possibility theory*.*

Incomplete information has been one of the curses of production management since line production and mass manufacturing began. Its manifestation is:

- Haphazard data gathering, little or no data validation,
- Ineffectual databasing,
- Inadequate data communications, and
- Obsolete methodologies.

Figure 12-2 brings into perspective the main characteristics of a valid methodology developed to respond to dynamic requirements on the factory floor. Most of the modules that constitute this solution use heuristic reasoning:

- *The input* consists of production plans, factory resources, order status, and so on.
- *The processes* taking place in scheduling optimization are essentially demand swappers.
- *The constraints* result from work in process, optimal sequencing, and labor stability.
- *The output* is resource reallocation handling both proactive and reactive schedules.

The main goal is that we are joining flexibility with consistence. Inflexibility leads to a state of affairs that negatively impacts on the production process. Since planning is a cornerstone of this approach, the proper information is not available to support precise

---

* See also D.N. Chorafas, *Knowledge Engineering*, Van Nostrand Reinhold, New York, 1990.

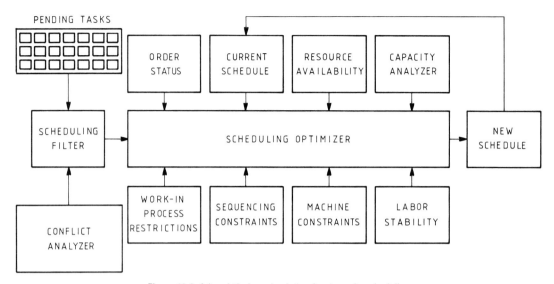

Figure 12-2. A heuristic-based solution for dynamic scheduling.

scheduling decisions. It lacks results and an ability to react to changes, which are seen too late to permit optimal rescheduling.

In the typical plant, the effects of incomplete information are magnified by a *divided production structure*. This sees to it that there is an uneven distribution of decision making splitting engineering and manufacturing, promoting watertight work centers, and reducing the coordination that should exist between tooling, materials, and robot programming. In short, there is a dispersed hierarchy that impacts negatively on:

- Cooperation,
- Coordination, and
- Resource utilization.

This situation is often aggravated by the prevailing *constraints*. They may be due to *causality* such as the precedence of operations and assemblies, or of resource requirements, i.e., machines, tools, jigs, fixtures, and most importantly, qualified personnel. They may also be due to physical capabilities and limitations of the machines or logical issues relating to machine programming.

Another major manifestation of constraints has to do with organization goals, for instance, meeting due dates, reducing work in process, swamping costs, equilibrating production levels, and doing away with ups and downs in labor employment. Much of this originates from preferences and operational choices, often having little to do with rational decisions.

A great deal of the challenge arises from the fact that such constraints are often ill defined and, in some cases, fairly complex. Valid solutions require the simultaneous satisfaction of multiple constraints in spite of a changing environment with possibly inconsistent or conflicting goals.

3. VITAL SUPPORTS IN PRODUCTION MANAGEMENT

Stock can be consumed only in the future. Decisions to manufacture must therefore be based on a forecast of future demand and this forecast should be frequently reviewed if control is to be effective. Basically, the production planning and control problem is concerned with the organization of men, machines and materials, with the objective of manufacturing a product or products:

- In a preestablished time,
- At a predetermined rate, and
- At the most economical cost.

Every production plan aims to utilize existing resources (the plant, labor, and finance) in the most profitable way. In this sense, the contributions of computer-processed mathematical models involve their ability to carry out the functions of production planning (and replanning), their introduction of efficiency into the control cycle; and their making provisions for automatic feedback.

Nevertheless, effective sales and production coordination can only be achieved when overall company objectives take precedence over local functional objectives. The control of production has only one—although important—effect on overall profitability. Hence, the plant manager should be aware of the overall corporate plans and he or she should work *synergistically* with the other departments rather than against them.

This principle of synergy sees to it that the process of planning production through computers starts at the sales forecast level. There is where processing specifications and production schedules originate, later accounting for the need to determine how to produce reliable quality items on schedule by using available resources.

Decisions relating to production management can be so much more effective if the available information technology significantly reduces clerical tasks and also improves data inflow and outflow from production engineering. This typically takes the form of sequential steps the way it is outlined in Figure 12-3. Let's notice, however, the significant impact feedback can have.

Work simplification is a vital part of an agile production planning methodology and it can have a significant aftermath. In one application I made a few years back, a dozen major records that had to be semimanually, semi-DP prepared and maintained at the plant site, were replaced by a single document that was available online:

- The system accepted process and source data which it analyzed and recorded.
- Networked workstations produced planning and ordering documents and provided work standards information.

In this particular application, data on cost evaluation were available online, realtime, based on standard costs. Ad hoc queries were acceptable and led to interactive performance softcopy reports which showed a comparison of estimated man-hours against hours actually worked.

This fully interactive approach covered not only the plan but also the change of production

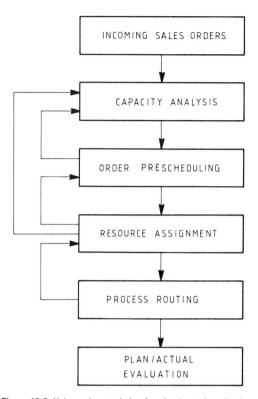

Figure 12-3. Using sales statistics for plan/actual evaluation.

schedules (and the resulting downtime or upset) because of the change in sales demands. Furthermore, on request, a variety of ad hoc reports were produced to answer specific questions pertaining to robots, tools, parts and man-hour requirements. We shall see how mathematical modeling and expert systems aid in this task.

The process of determining the number of parts required to build a planned number of given products is a *requirements generation* and closely relates to the Bill of Materials to be prepared at the design phase. It is the job of both the designer and the manufacturing engineer to:

- Examine the nature of the records necessary to do a requirements generation,
- Investigate and elaborate a flexible but secure method for such a generation, and
- Establish the ways in which the results of a requirements generation may be altered.

Particular attention should be paid to the record of the parts that compromise an assembly description, that is, the Bill of Materials. As we have seen in the preceding chapters, this is a list of parts required to form a finished product or an assembly.

Generated at the level of CAD, the Bill of Materials must have three basic entries for every part listed:

- The part designation,
- The internal or external source of supply, and
- The quantity required.

Other data may also be needed. When the structure of a manufactured product requires that subassemblies be noted, then several Bills of Materials are combined to describe the product. The structure of the product is then described by a hierarchical nest of such bills, each of which represents an assembly or subassembly.

Computer-run Bills of Materials logically map the product into the machine and can be used to develop, level by level, the materials structure of a given product. This structure is expressed in terms of reasonably simple relationships that are of value to cost accounting, engineering and production planning. They also serve in inventory control, considering parts in stock or in order.

Purchasing and inventory control decisions depend on the information that has just been described. Such a logical organizational structure helps guide the production planner's hand in handling operations for multiple orders while he or she obeys the temporal restrictions of production processes and the capacity limitations of a set of shared resources, thus balancing conflicting sets of requirements.

A modern production planning system should reflect the work currently done using classical approaches but it should also use expert systems to convert all data into a production plan according to:

- The general scheduling policy,
- The production sequence and flow, and
- The current manufacturing status of each article in process.

By mapping into the computer the skills and knowhow of expert production planners, schedules are documented on a controllable basis including active data for ordering, manufacturing, stocking, and issuing items. Requirements for materials, machinery, tools, labor, and performance measurement standards can be obtained through interactive reports along ad hoc or structured levels of reference.

An *enterprise level* will reflect planning for an aggregate resource; it will feature a time horizon dependent upon available machinery and associated lead times. A *factory level* will focus on detailed scheduling including job sequencing, order release dates and activity coordination. A floor level will address itself to dispatching, including realtime job-machine allocation.

Manufacturing management should be actively called to take a part in this process by evaluating intermediate results prepared through simulators and expert systems:

- If the comparison is satisfactory, the detailed production plans will proceed in execution, including shop order data, operation inspection logs, and make-or-buy authorizations.

- If the AI constructs detect an error in decision, corrective action should be taken. Part of this action may result in the need to review and revise their guidelines.

In summary, then, computer-based models will perform the thousands of steps required to check and file the production planning guidelines, resource identification, order flow, and optimization perspectives as well as the production budget, including delay, downtime or quality penalties. Simulators and expert systems will optimize the manufacturing process, plan the use of the production resources and establish an effective communications channel with management—from the headquarters to the production floor.

4. MATHEMATICAL PROGRAMMING AND SCHEDULING METHODS

From a theoretical production management viewpoint, scheduling methods are available to deal with the issues we have been discussing. Their practical availability is, however, often compromised because of the obsolete thinking in production planning situations. Many, if not the majority, of approaches taken today in the scheduling of production are imbedded in outmoded ways of thinking. In other cases, there are rundown equipment and inadequate manufacturing methods, their faultiness being compounded by the increasing complexity in fabrication methods and schedules.

Clear-eyed manufacturing executives realize that they need adequate tools to overcome this complexity. That is why they use mathematical programming and expert systems.

For nearly four decades, one of the best problem-solving methods for scheduling problems has been linear programming. But production planning problems cannot be solved directly via *simplex*. At various times other approaches have been tried such as dynamic programming, search and prune, as well as linear programming relaxation.

As the name implies, with linear programming relaxation, the criterion is to relax the integer constraint that contradicts the use of the simplex method. This creates a linear programming background that is easier to solve at least at the lower level.

Another approach is the so-called seed solution which uses linear relaxation as a starting point for branching: A hierarchical decision tree is developed with a number of leaves. The solution lies in:

- Proving that parts of the tree are inferior and then
- Pruning these parts away until exhausting the tree leaves are exhausted.

The constraint associated with this approach is that its complexity increases exponentially with the size of the problem.

Another problem-solving method in mathematical programming is Lagrangian relaxation. It consists of moving some integer constraints into the objective function using Lagrangian multipliers. The resulting integer program is supposed to be easier to solve, yet it is not so much easier. Also, since some constraints are missing, the solution is the lower level for any value that may result from experimentation.

This brief list of approaches to mathematical programming is not all inclusive but it does give us an idea about what we are up to when we intend to apply a sophisticated scheduling methodology. The fundamentals are fairly simple, consisting of identifying and focusing on the simpler possible reformulation of a relatively complex problem. The implementation is, however, much more involved. To do a relatively simple reformulation:

- We model the problem quantitatively,
- Account for its qualitative components,
- Try to transform the problem structure from a more complex one to a relatively simpler one, and
- After successfully mapping the problem into the computer we select an optimization method.

Through experimentation we determine the solution to the simpler problem, as an approximation to the more complex one. If such a solution is not close enough, then we restructure our model.

This approach is being applied with increasing frequency in the manufacturing industry, and production planners have come to the conclusion that a major drawback often lies in the almost exclusive focus on *quantification*. For almost 30 years (from the mid-1950s to the mid-1980s), quantitative modeling has been the primary technique, whether it is linear or nonlinear.

Constraint satisfaction has been a primary tool for problem solving and linear programming is the approach most used. Yet, little by little, we learned that we cannot handle the complexity of a good deal of factory scheduling problems this way. As a result, since the mid-1980s expert systems have been used adding *qualitative* decision factors to the quantitative approaches through the use of heuristics.

Expert systems assist in establishing an *object-oriented* representation of the factory. They bring into being a pattern-directed inference in the form of rules, frames, or scenarios. They emulate production planning experts, simplifying their approach by decomposing the problem into separate, though interelated, tasks.

Through the use of AI constructs, each production planning task is solved using rules extracted from the domain expert himself. Then the expert system employs the rules applied to the scheduling or dispatching jobs. This is a good starting point for the significant issues a scheduling system should attend to—and there is no reason why such expertise cannot evolve over time.

In one specific implementation, the knowledgebank is designed to minimize the set-up time required to retool machines that produce products of different stocks or sizes. Variable set-up times are stored in the database. The expert system reads two files as input:

- One file contains the work orders that need to be scheduled.
- The other contains the state of all machines and products for which they are tooled.

As output, the application produces a work schedule and a list of work orders. The following approach is used to schedule them:

- Choose the order to be scheduled based on due date and size, giving preference to the one that has the earliest due date and largest size.
- Try to schedule the work order on a machine aggregate already tooled for the stock specified in the order.

- If unsuccessful, try to schedule the work order on a single aggregate regardless of what stock it is tooled for; alternatively, split the order between multiple machines.
- If scheduling the work order is still unsuccessful, postpone it.

This application takes advantage of object-processing features such as inheritance, specialization, demons and pattern-matching rules. Full message passing has also been used, permitting the developer to:

- Send messages with arguments to objects,
- Write methods that return values, and
- Control when messages are passed.

Data is represented using objects, and pattern-matching rules have been used to implement the scheduling algorithm. Object processing allowed for a clearer specification of the algorithm, improving both readability and maintainability.

One of the more successful implementations of expert systems in production scheduling is constraint-directed search. The representation is done through semantic networks that define factory planning knowledge as well as constraints, whether exogenous or endogenous, both being elaborated to include relaxations.

Such constraints define operators and help in terms of a constraints-bound search. This helps in specifying decisions by defining the evaluation functions to be used. Typically, problem formulation focuses on a contention over resources, which is at the heart of every planning problem, production scheduling being no exception.

5. THE ASSIGNMENT OF RESPONSIBILITIES

Within the perspective of the mathematical programming approaches and expert systems assistance discussed in section 4, the ABC Manufacturing Corporation saw to it that a methodology was developed calling for an in-depth analysis of the current production planning methods, which were able to integrate into one system sales orders and scheduling. This can be structured into a four-phase approach:*

1. *Promises*—essentially obligations being taken as the due date and other delivery characteristics.
2. *Plan,* which focused on capacity, orders, resources, and availability.
3. *Scheduling* proper with the starting date and subsequent queueing as well as inventorying—using heuristics.
4. *Assignments,* that is, priorities leading to and establishing the release date.

The promised date (due date) is an external constraint. Planning premises are internal constraints. *Slack* used in scheduling is a heuristic. An example of this is the maximum slack rule.

---

* This four-phase presentation is based on a discussion with Dr. Alan Rowe of the University of Southern California.

The study started where it should; it evaluated the strengths and weaknesses of the current approach. The project team was held responsible for defining what was supposed to be done; how people thought it was being done; how it was actually done; and then, how it could best be done.

The products this company sells address two different markets: a consumers' market (product lines I and II) and an order type manufacturing market (product line III). Each product line required special attention. However, prerequisites by product line brought conflicting forces to bear on production and inventory planning; hence, an integrative approach was taken.

Figure 12-4 identifies the overall concept that was elaborated from user data to the planning problem solver, and it was to be the kernel of the solution. The following objectives were established for the production planning system to be designed with the assistance of mathematical programming and AI constructs, such objectives focusing on the need to improve the company's competitive position:

- Significant improvement of customer service
- Better production analysis capabilities
- Flexibility in production scheduling
- Development of accurate, automated cost control procedures
- Reduction of inventory accumulation and avoidance of its obsolescence

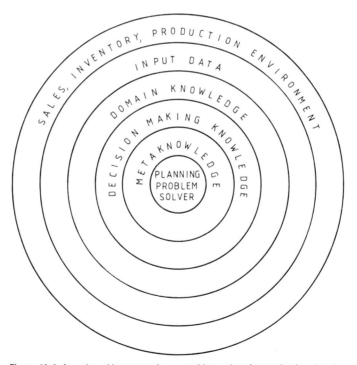

Figure 12-4. An onion-skin approach to a problem solver for production planning.

- Improvement of man-power planning as well as facilities planning
- Increase in office effectiveness with the associated reduction of clerical costs
- Development of more efficient methods to modify and disseminate production-planning information.

A basic concept of system implementation required that jobs having the greatest effect on profits were handled first. In accordance with this, one of the first jobs to be tackled was the control of materials, specifically parts and finished-goods inventories.

A strong emphasis on producing just-in-time finished goods and components for stock was present in this endeavor. The traditional production management solution had been to stock some items fully on the finished goods level, to stock others partially, and to leave still other items on the component or raw material level, producing them only when orders were received. But inventory optimization was not part of these goals, the result being time series, as seen in Figure 12-5.

If we leave for a moment the issue of overstocking, such an arrangement of production/sales coordination provided an adequate approach to the problem presented to this company for a number of years, given the timing constraints. But as customer demands became more complex it failed to respond to them in a flexible manner. As a result, a group of expert systems was developed each in a modular way capable of integration. These AI modules addressed themselves to:

- Determining and allocating the material, labor, and equipment needed.
- Controlling inventories of finished goods, components, and raw materials.
- Providing input data for scheduling, production, and delivery.

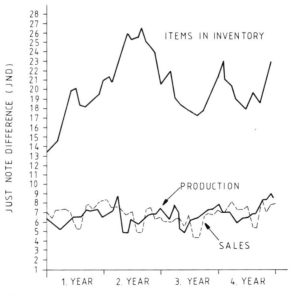

Figure 12-5. Sales inventory and production statistics of a given product over a four-year period.

- Handling the factory assembly and fabricating orders, as well as purchase requests.
- Optimizing equipment and labor usage including interactive reports on labor, machine time and material expended.
- Calculating direct and indirect costs, following standard cost criteria needed to establish conformity to market prices and to maintain profits.
- Supplying accurate and timely exception reports.

The results of implementing some of these expert systems make interesting reading. The modules addressing themselves to scheduling identified some *gross inefficiencies* in the production floor. These were swiftly corrected but the result has been that the scheduling rules changed and the originally designed planning construct was no more real. Sustenance had to be considered.

Key to the effective integration of the many separate modules of this solution was their networking. This, as well, included realtime access to databases that contained basic information about the business. In the design and implementation of this approach, particular attention was paid to the personal responsibility of sales, inventory and manufacturing executives of the firm, pointing to personal accountability.

In a way, the concept of management accountability and efficient mathematical programming solutions assisted one another:

- The responsibility of the factory manager for costs and profits will remain an empty letter until and unless the appropriate tools are given to him to monitor the production budget interactively and to maintain an efficient use of resources.
- Without interactive capabilities, even the most carefully established and updated production planning or standard costs scheme cannot be used as a means for judging production management capabilities.

With or without mathematical programming and expert systems, the smoothing and optimizing of production will *not* be painless. The key word is *change*. A profit-minded organization can accept no transmutation of old practices into different modes. The computer will not improve ossified approaches.

It is always wise to keep in mind that when it comes to the solution of scheduling problems, companies need to have imaginative approaches to remain competitive. Contrary to classical procedural routines, knowledge-based approaches employ the same heuristics used by experts and find reasonably good schedules relatively quickly.

Such approaches are more adaptable and thus less brittle. Furthermore, a knowledge-based system can explain its reasoning and produce partial solutions if an optimal one is computationally infeasible or cannot be found.

Production efficiency calls for new departures. To turn this wish into reality, a full integration from sales to inventory to production should come into effect stressing the capability of the total system. An able solution would start at the point of origin (data collection) and end at the execution of the customer's order, passing through production planning and control and also including inventory management.

Unless full coordination is achieved in that sense, it will not be possible to face the requirements of an optimal production plan. If full coordination fails because of certain departments keeping the needs of the neighbor departments within the company operating process watertight, then production planning will not be achieved in an optimal, profit-oriented sense.

6. IMPROVING THE OVERALL EFFICIENCY THROUGH INPUT/OUTPUT SOLUTIONS

In the case of ABC Corporation, it was decided that daily control and feedback will be executed on the basis of daily detailed schedules. Based on a sound plan elaborated through mathematical programming and expert systems, the foreman will control the assignment of machines, men and materials, as well as the movement of tools. As each assignment or move is made or completed, the expert system is immediately notified with both qualitative and quantitative information incorporated in such notification.

This reference brings into perspective the need for able solutions regarding to input/output (I/O) facilities. A list of the more important classes of input/output functionality was therefore compiled and extended to include the auxiliary services normally needed for convenient and efficient operation.

The provision of efficient input/output facilities is an issue that has been classically the laggard of the computer business. Since the 1950s the design of the first group of electronic computers tended to emphasize the development and exploitation of high-speed arithmetics and logic circuitry. The *input to* and *output from* this circuitry were improvised, using the equipment available in the telegraphic communication field and the punch card business machines.

Neither of these two fields of carry-over concepts and equipment happened to be of any great service to manufacturing. Stated simply, they were not made to fit the requirements of a fabrication floor, and this is true even if in the 1950s and 1960s:

- The traffic into and out of the computer for the class of problems under consideration was quite small compared to the internal operations.
- Because the knowledge about how to use computers effectively in production was thin, the lack of balance between input, processing and output was not an excessive burden to the user.
- Experience accumulated, and it became apparent that many important tasks to which computers were applicable, in principle, were not practically handled because of the limitations imposed by the retrograde input/output.

If certain types of scientific problems could not be processed in a completely automatic manner unless there was some means of circumventing the limitation of storage in internal memory, the input/output limitations saw to it that certain areas of statistics, tabulating, accounting, and related record-keeping tasks required a far greater sophistication in I/O.

This was even more pronounced in scheduling—in balancing tasks, time slots and resources.

Today, years of experience help document that a flexible high-performance, fully interactive input/output facility is the necessary support to production planning and control:

- The primary function of an agile input/output media is to capitalize on the time scale of the data flow between the relatively slow external world and the very fast internal world of a computer.
- A second function is to provide a method for experimenting on active records between processing runs with an adjunct interactive presentation facility.

Resource allocation is a scheduling problem that requires good coordination to be handled in an able manner. Valid functions can be attained through the appropriate organization of input/output, particularly if we use modern means such as:

- Natural language comprehension,
- Voice recognition,
- Image processing,
- Object modeling,
- Pattern recognition, and
- Artificial vision.

This means that we must fully master the foremost areas of artificial intelligence. Provided the right input is assured in a timely manner, AI constructs have been successfully used to identify images and sounds. Artificial vision is basic in guiding robots and in processes of surveillance and quality control.

The importance of this reference will be better appreciated if we keep in mind that, as underlined in Chapter 8, in a modern factory robots are composite entities that may be:

- *Stationary,* such as assembly-line arms, or
- *Mobile* and designed for autonomous tanks.

Input/output perspective and modern technology merge together into robotics, as a robot typically integrates systems from varied AI disciplines: vision, natural language, problem solving, expert systems behavior, and planning. An application of hypothetical reasoning is a planning application, for example, the type of plans that a robot may have to construct in order to accomplish a desired end.

A different way of looking at this same subject is that production planning issues both require and impose communication and cooperation between subsystems, more demanding than those that would involve only solving a problem alone. A good example is provided by robot intelligence which greatly depends on I/O.

Indeed, robot intelligence is one of the most attractive and of the least clearly defined subjects. Psychologists and biologists study perception, cognition, learning, and other aspects of intelligent behavior in natural organisms—with *input* being the cornerstone of each one of these subjects. Research on input:

- Ranges from humans down to the most primitive animals and insects.
- And goes all the way up to the design of autonomous vehicles.

We no more live at a time of self-congratulation when we demonstrate that a variety of peripheral equipment is available as auxiliaries to a central data processing facility. Though still used in large numbers, keyboards, output typewriters and line printers are not valid I/O. The nature and volume of the input/output work has changed. Magnetic tapes are today the worst solution to be found for input, and the same is true of hard copy output.

Systems solutions in production control require both a convenient and effective means for transcribing source information from a given form onto input facilities. I have mentioned voice input; other solutions are facsimile, scanners and curve following devices coupled with analog-to-digital converters, as well as the more recently developed reading devices for handwritten information using neural networks.

All this is highly relevant to the production planning landscape as robots and robotic systems are becoming vital performing elements in many manufacturing environments. It is important to know not only what robots can do but also how to improve their intelligence as well as their I/O. Only then can we plan their utilization in a valid manner.

7. DISTRIBUTED NEGOTIATIONS FOR SALIENT PROBLEMS IN PRODUCTION SCHEDULING

A salient problem is one to which management will address itself first, on a priority basis. In scheduling, this may be the case with representing work sequences and reasoning about time; establishing causal relationships between actions; facing uncertainty in the execution of plans; and assuring that multiple agents cooperate and do not interfere.

There are physical and logical constraints on suitable solutions to salient problems and many of the knowledge-based planning techniques seek to control the extent of the search for valid planning premises. This presupposes that the input/output problems have been efficiently solved and that the information in the manufacturing database will always be fully current with the progress of the physical processes being controlled.

The concept of integrative solutions is that in a manner similar to the way in which it generates manufacturing requirements, the information system is expected to:

- Convert them into purchase orders, and
- Follow up with an inspection of the materials being received.

Purchase order information and appropriate inspection feedback is an integral part of any manufacturing database, as we will see in Chapters 15 and 16 when we talk of computer-integrated manufacturing.

At the ABC Corporation, the approval of vendor invoices, preparation of vendor payments, full accounting procedures, inventory control, and associated management reports were executed as coproducts of the purchase order procedure. This was seen as a major improvement in efficiency and in the control of operating cost, most particularly in critical evaluations and analyses.

To obtain this goal, expert system modules were written to help in *routing*—from the determination of the route for the movement of a manufactured item through work posts in the factory to that concerning external suppliers. In both cases, the AI constructs help define the status of each operation on a component, subassembly, or assembly.

Let's add, however, that in production engineering, routing is a term with many shades of meaning, depending on the firm. While nearly everywhere it relates to the process of deciding which operations are to be performed to make a part, including the type of machine and machine attachments needed and the sequence of operations, it also often refers to follow up on the status of parts and operating lists, or to copying them into production orders.

Another mathematical programming module addressed itself to *loading*, including the distribution of the required work load against the selected machines or workstations. This involved algorithms rather than heuristics: The total time required to perform the operation is computed by multiplying the unit operation time by the number of parts to be processed, resulting in an interactive chart that shows the planned utilization of the machines and workstations in the plant.

A detailed *scheduling* expert system helped determine when an operation is to be performed and when work is to be completed. When all routing, loading and scheduling for the plant is done through computer-based models, reporting from the ground floor must be executed in realtime to ascertain whether the execution of the schedule is in control.

In a way similar to the DEC implementation we saw in Chapter 11, *dispatching* AI modules refer to the issuance of orders to produce as well as to expedite. They assist in:

- Issuing instructions to the fabricating department(s),
- Authorizing the start of an operation on the shop floor,
- Controlling the execution of each ordered operation,
- Registering the results of quality control, and
- Assuring the proper expediting procedures.

All this conforms to the concepts that were highlighted at the end of Chapter 11. However, what makes this particular implementation different is the process of *distributed negotiation*.

With the AI constructs that have just been described as networked, a node that has a job to be done broadcasts this fact to all other nodes, and the other nodes make bids. This is facilitated by the fact that the knowledge engineers who designed this solution have carefully constructed a simulator of the factory, making it available online.

Such a process is enriched by a discrete event simulation: When a job arrives or a machine becomes free, local dispatch rules are used to select the next job to load. The expert system optimizes this process with two rules:

- The shortest processing time, and
- The earliest due date.

The simulation is followed by an optimization which results in a definite schedule—after having brought to the attention of the production planners conflicts that might exist in local versus global requirements.

Conflicts of this and a similar nature get back to some fundamental issues concerning production planning:

- Precision versus imprecision,
- Certainty versus uncertainty, and
- Predictive (proactive) versus reactive solutions.

Such conflicts are resolved through *constraints* and, as is known from other experiences, constraints are necessary to provide problem boundaries.

A good example of constraint implementation is *rescheduling* done in sequence to corrective action. As Chapter 11 underlined, it consists of reviewing routes, loads, and schedules. Expert systems help a new plan develop. The key issue, however, remains:

- How to make a local optimization globally valid, and
- How to satisfy global requirements in terms of production schedules.

A distributed negotiation environment makes steady use of constraints. Its representation is done through expert systems that act as separate agents distributed across the production planning landscape. Each agent manages its own tasks and an operational contract defines the relationship among them. With this, the problem is reduced to one of a protocol of negotiation:

- A factory supervising agent announces task(s) for a given job, most particularly the task's first operation.
- The supervisory agent's call defines the start, duration, end time and resources, with the starting time contained within a range.
- Work center agents respond with a bid that defines the cost and the date to deliver.

Having received such bids on line, real time, the factory supervising agent will award a contract that is bid on and that meets due date and cost requirements. It will then send out a call for a second operation. Each contractor agent bids on the calls and if successful it is awarded a contract.

Another expert system the ABC Corporation found to be quite valuable is *data filtering*. Its importance is the direct result of the fact that in real-life situations production scheduling has two main problems located in the antipodes:

- Too much data, and
- Too little data.

Both are detrimental to the elaboration of valid approaches to a production-planning solution; hence, we see the wisdom of applying appropriate filtering solutions.

From distributed negotiations to data filtering, the solutions we have considered are particularly appropriate when the factory planning environment is fuzzy due to market conditions or other reasons. The net result of the advised approach is planning simplification which, after all, is the role of simulation and expert systems in connection to scheduling premises.

13

The Challenge of Quality Assurance

1. INTRODUCTION

Although scheduling for optimal utilization of resources is a complex and demanding task, the most critical problem in automated production is that of *quality control* (QC) and product testing. In order to remain competitive, it is a common practice in industry to develop methods for checking the quality of materials and components in process quickly and automatically and to test the performance of the completed product.

The difficulty of the problem lies not so much in the building of automatic test or control equipment, but in the conceptual and organizational side:

- Determining what physical characteristics are to be measured to provide a key to the reliability of performance of the component or the assembly,
- Establishing automatic testing procedures proper to robotic solutions, which require artificial vision and powerful quality detection algorithms.
- Building and running a quality data base enriched with heuristic modules, able to explore its contents, leading to corrective action.

One of the most difficult parts of thinking in terms of automated production is recognition of the fact that the product as part of the fabrication process must be designed in order to make quality control automation feasible.

We must look at the function of a product and think about building it and inspecting it by automatic machinery. We must also rethink our organizational solutions, as Chapter 12 underlined for a number of vital reasons involving production planning and scheduling.

What are the functions a modern, dynamic quality control department should perform? At Hughes Aircraft Company, the Quality Information System (QIS) department compiles and analyzes data on how defects happen and how they are corrected:

- Information is made available to manufacturing employees for immediate feedback and use during production.
- Data is also kept in a central historical (quality database) for future analysis and reference.

QIS improves quality by spotting problems that stem from faulty design, poor supplier quality, and improper manufacturing methods. In addition, by keeping track of how often certain hardware needs replacing, the system helps to pinpoint problems in the production line.

A cornerstone of the proper functioning of a quality assurance operation is the ability to measure, monitor, improve and control quality through the availability of reliable production and field information. Also important is the ability to analyze this data statistically on a real-time basis and to make comparisons, execute tests, and reach decisions.

In modern engineering, we devote more effort to the study of new materials and apply fruitful thinking unencumbered by conventional design practices. Therefore, it is not beyond the realm of reality that in the not too distant future we shall be able to produce products that will perform reliably by more than an order of magnitude as compared to current products.

Seen in a statistical distribution sense, high quality means a very small standards deviation around the mean value of the product's operating life, which should improve steadily. This beefing up of quality now needs to be done in a production line practically untouched by human hands. In these few words lie the quality challenge of the 1990s and beyond.

2. GAINING LEADERSHIP IN QUALITY ASSURANCE

One of the most important domains the Japanese industry implements in its expert systems is its quality control applications. The aim is to structure expert knowledge in a manner able to integrate the knowhow of several persons, enriching the factory's quality perspectives, and distributing QC knowledge within the organization.

This policy seems to bring commendable results. In September 1990, the Massachusetts Institute of Technology unwrapped a follow-up to its *Made in America* study. This new volume: The Machine that Changed the World, focuses on the American, European, and Japanese auto industries.

The data contained in Table 13-1 stunned senior American executives. The reason is the disparities that exist between America and Japan. As Business Week was to suggest,* MIT researchers were surprised that US automakers had not done *technology audits* to learn why their foreign rivals are so successful. But what really astounded the MIT team was:

- The lack of knowledge in the financial community and Washington
- About the fundamentally different economic solutions that Japan has brought to world competition.

---

* October 8, 1990.

TABLE 13-1 MIT Findings on Quality and Productivity Data: Japanese versus American Practices in Auto Production

	Japanese plants in North America	U.S. plants
	Hours	
Defects per 100 cars	65	82
Assembly time per car	21	25
Worker training	370	46
	Days	
Inventory supply	1.6	2.9
Work teams (Percent of work force)	71%	17%

Yet it was the American know-how that developed the necessary infrastructure for dependable quality control procedures. The father of modern quality control has been the statistician Walter Shewhart who, in 1924, working at Bell Telephone Laboratories, applied statistical tools to reduce defects. The Shewhart QC chart has been a milestone ever since.

During World War II the operating characteristics (OC) curves came out of research done in connection with the Manhattan Project at Columbia University. After the war years, consultant W. Edwards Deming—ignored at home—lectured top Japanese executives on statistical quality control in manufacturing and became a quality guru in Japan.

As usual, the Japanese improved upon what they learned. By 1960, Toyota started applying the very first steps of the design process, the quality assurance methods developed by the engineering specialist Genichi Taguchi. These were turned into firm policy improving quality at *the design stage*.

It is also the Japanese factories that initiated quality circles, effectively teaching production workers ways to cut defects and solving a number of other problems through this process. This policy of quality circles was then extended into service sectors, such as the banking industry.

What Genichi Taguchi, W. Edwards Deming, and Walter Shewhart have achieved in quality assurance, is fully applicable to every company—all the way from design to manufacturing and maintenance. Quite definitely we should capitalize on this pool of know-how which has been tested over the years and is today highly appreciated.

At the same time, mathematical methods can make major contributions in quality control as well as in controlling the variability of causes behind defects. Figure 13-1 comes from computer manufacturing and identifies missing wires in the background panel.

A sample of 10 days is displayed in the histogram with a distinction made between wrong connections and other causes. As it can be seen, in this small random sample, wrong connections vary between 2 per day and 30 per day. There are significant advantages to be gained in terms of higher quality and uniform production through statistical quality control.

The attention devoted by the Japanese in the matter of quality assurance pays its dividend. In a recent *Business Week*/Harris Poll involving 1,253 adults, 77 percent of those queried drove US models, but:

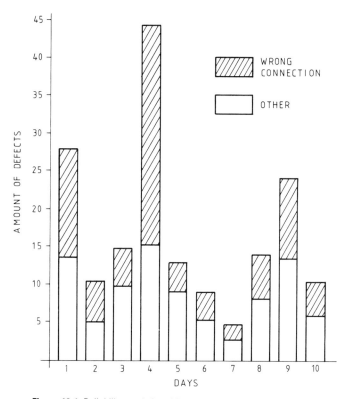

Figure 13-1. Reliability statistics: Missing wire on background panel.

- Only 61 percent said American cars are better than those from Japan or Europe, and
- While 96 percent consider reliability very important, only 46 percent thought US cars have fewer defects than Japanese cars do.

Most significantly, in the context of these research findings, only 51 percent said American cars are better engineered and superior in every way.

No wonder US auto manufacturers are now actively searching for solutions supportive of product quality. But in a fully automated environment such solutions do not come easily unless they start at the darfting board, and a whole system is built around them.

As *Business Week* reported,* when Buick City started up in September 1985, it experimented with a just-in-time inventory. Hence, orders went out to nearly 600 suppliers on:

- How many parts to ship,
- At what time of day, and
- In what order.

---

* October 22, 1990.

The schedule was calculated to match cars coming down the line. With only an hour's worth of parts on hand at times, defects often shut down Buick City's production line.

Suppliers were sending better parts than they had ever sent before, but they were not good enough for such an automated system. Problems also developed in Buick City's technology, particularly in its 250 robots. For example, artificial vision devices on robots that installed windshields could not see black cars, so they skipped them until new software solved this problem.

There were consequences, however. As a result of restructuring, a low two-digit number of the plant's robots were mothballed. Furthermore, to reduce costs, GM eliminated production-line inspectors and cut the number of supervisors in the plant in half:

- The job of assuring quality was given to assembly workers.
- Each worker individually could halt production by pulling a cord at his workstation.

By late 1986, the plant was in trouble. An estimated 200 LeSabres produced as a test run for the '87 model launch were too shoddy to ship. Engineers worked on the line to discover firsthand the bugs in parts they had designed, and the "stop the production line" system was modified to include a warning cord so that workers could call for help without stopping the line.

In the quality control front, since Buick City opened, some 80 suppliers have been dumped because of not meeting the established quality goals. However, in 1989, the Oldsmobile Eighty-Eight Royale made at the plant had 130 problems per 100 produced, 59 percent more than the Buick LeSabres—which talks volumes of quality problems embedded at the design stage.

The issue here is not that the Japanese are doing everything better in the quality assurance front. Rather, the underlying issue is that we should be eager to learn from what others have achieved, revamp it for our own purposes and do better than they did. A precise application of this type of reasoning involves the use of knowledge engineering in quality assurance.

3. ARTIFICIAL INTELLIGENCE IN QUALITY CONTROL

Expertise embedded in AI constructs can help unearth the underlying causes of a wasteful production process, reducing the cost of defective and reworked parts, and scrap, and contribute toward the manufacturing of a more uniform product. Operating on line, quality control expert systems make much better use of engineering tolerances and help obtain a significant level of quality assurance.

Experience along this line of reasoning started in the World War II years with mathematical statistics. The use of statistical quality control programs decreased inspection costs and provided for a more efficient use of materials, as well as the establishment of economical inspection procedures. Other results included improved and concise quality reports to management, and more effective relationships with vendors and consumers, that is, business partners.

These experiences provide the impetus for the implementation of a more sophisticated methodology made feasible through expert systems. The necessary information for the preparation of a quality picture include data:

- Received from suppliers of materials and components,
- Collected by the inspection department in exercising its duties of inspection (on materials, parts, and final products), and
- Obtained from the maintenance engineers as well as from customers with respect to failures in the field.

Although it is important to know the quality of the finished product as it leaves the plant and the caliber of the quality job being done by the inspection department, it is just as vital to follow the product in its field usage. This process can be automated through online diagnostics as we will see in the subsequent sections of this chapter.

The implementation of expert systems in quality assurance is particularly useful when:

- Few experts are available for quality inspection,
- The quality inspection process is widely diffused,
- A special knowledge is necessary, particularly in exception handling,
- A certain transparency of the quality assurance solution exists,
- There are too many or too few consultations on quality matters,
- The obtained quality knowledge is broadly distributed, and
- The expert system is also used for training purposes.

In section 2 we saw examples with Buick City suggesting how much more effective quality inspection would have been if expert systems had been used online in the assembly process. Apart from dealing with quality assurance problems per se, in several cases AI constructs provide recommendations regarding maintenance procedures for the prevention and repair of production facility malfunctions.

An example is given in Figure 13-2 where an expert system identified wasted time by shift, then provided recommendations for process adjustments. This was achieved by properly identifying where and why manufacturing process deviations occurred.

In this, as in a number of other cases, parametric control took care of reducing quality problems. Sometimes the impetus for such implementation comes through an audit that reveals weaknesses in terms of quality observance.

In the early 1980s, an internal audit shocked Westinghouse officials by revealing that 88 percent of the printed circuit boards (PCB) the company was producing for aerospace and defense applications required some corrective work before they passed the final test stage. It was small consolation when a subsequent survey revealed that other defense suppliers were experiencing an even higher rate of rework.

Armed with this audit's results, the company made a commitment to improving the first-pass yields of its board facility. After undertaking a major automation effort that cost more than $50 million, and introducing innovative employee-relations practices, Westinghouse boosted the first-pass yields:

- From the previous 12 percent
- To more than 80 percent for mature systems.

THE CHALLENGE OF QUALITY ASSURANCE

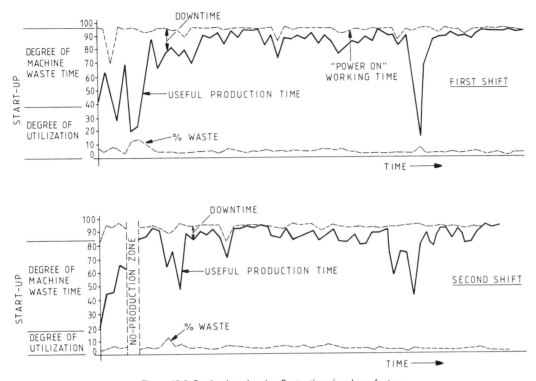

Figure 13-2. Production planning fluctuations in a lamp factory.

The long-term objective is to achieve 95 percent yields. This is an example of an area where expert systems solutions can be highly beneficial.

Developed by a European computer manufacturer, an expert system addresses itself to disk head assembly processes. Figure 13-3 shows the shop procedures and associated statistics.

- The manual evaluation at the pretester level indicates that 80 percent of the product is good; the balance has problems.
- The expert system has not been written to attack all of these problems but only the majority of them, which corresponds to the simpler ones.

The criterion for selecting this 16 percent of all disk heads (which essentially means 80 percent of those unacceptable) has been that the company employs expert quality control technicians who can solve the associated errors. This was the part converted to an AI construct as there were available:

- The knowhow, and
- Rules.

Such a procedure and the selection criteria it used can be widely applicable. For instance, in consumer electronics, boards come in relatively few types and are produced in very large

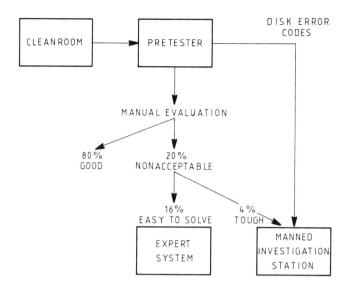

Figure 13-3. Control of a disk-head assembly process.

quantities, permitting quality controllers to gain skill in testing one single design. Such skill can easily be expressed in rules.

The opposite viewpoint, too, can lead to a valid argument though it will require a different type of approach to expert system development. The aerospace/defense field is characterized by relatively small production runs, hence:

- A workstation operator may produce only a dozen modules of one type before shifting to another, and
- Since size and weight are often critical in defense applications, the printed circuit boards are densely packed, increasing the risk of human error.

An investigation in terms of the rules that should command the control over quality, led to mapping into the knowledgebank the whole QC methodology as well as its decision criteria and procedures. Working in tandem with a simulator, an expert system has been instrumental in terms of quality upgrading helping to increase the outgoing quality level significantly.

Results can be further improved if proper coordination is provided with computer-aided design to expedite the transition from new-system development to production. Factorywide coordination through networks, workstations and expert systems helps engineers in designing products that fit better the constraints imposed by automated manufacturing. This statement is also true of quality assurance.

4. EXPERT SYSTEMS AS FAULT FINDERS

Today an increasing number of companies design expert systems to audit the design of their new products and to flash out conceptual and implementation faults. Typically these knowledge engineering constructs work within the CAD framework, with their implementation domain bridging the passage from computer-aided design to computer-aided manufacturing.

Applications of that type are custom made and this is a significant evolution from early AI research which concentrated on general-purpose problem-solving methods that could be applied uniformly to a wide range of applications. Contrary to such policies:

- The key word today is *focus,* and
- Our strategy is one of finding ways to use knowledge in well-defined domains.

The goal is one of successfully embedding superior knowhow into an operational system and quality control is an example of this type of application, provided we are able to describe the process, the product and the quality assurance rules in a well-ordered sense.

Figure 13-4 shows a diagnostic task model developed according to the foregoing principle. The operations performed by an expert system may be divided into a number of task types, such as diagnostics, design and control. These task types can be nicely represented using simple models that are clustered together in a comprehensive aggregate.

By following this process we are able to structure quality control expert systems where the degree of freedom from constraints (specified through partially ordered actions) permits experimentation with alternatives in order to make sensible selections:

- In production planning, time constraints specifying when the action should take place, make it feasible to tune the performance of a process in connection with the schedule being established.
- In a quality control environment, a set of goals can be given and expanded to greater detail. This will cover specifications as well as methods of restricting the allowable action ordering to one that could promote quality goals.

Hence, there is both similarity and difference in terms of these two processes. In essence, there are two ways to use AI in quality control:

1. Automate the control process by integrating it with CAD as we saw in section 2, and
2. Assist the human quality controller with online advice available to him interactively.

Parametric approaches can be thought of as a midground between these two references. In this way the process of designing a product will be merged with creative design and with quality inspection activities in a CAD/CAM sense.

At least one expert system attempts to integrate these two aspects into one well-knit solution. Known as the *function fault* model, this is an AI construct with associated

Figure 13-4. A diagnostic model for quality inspection.

222 CAD, SCHEDULING, MANAGEMENT AND CIM

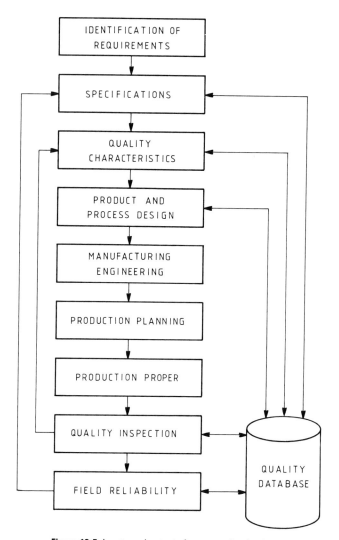

Figure 13-5. Inputs and outputs from a quality database.

problems solving methodology for technical diagnosis. Its first major component incorporates:

- Functions and subfunctions,
- Definition prerequisites,
- Description of components, and
- State values.

Correspondingly, its fault location modules focus on the:

- Interpretation of conditions,
- Identification of needed tests,

- Possible environmental effects, and
- Associated explanations.

Taken together, these expert system modules lead to problem identification capabilities by using an approach based on heuristic search and online access to quality databases.

This and similar examples help demonstrate that the creativity of the inspection job rests not only on imagination but also on databases. As a process quality, assurance can only be handled online through interactivity and visualization, going all the way from requirements description and product decomposition to analysis, planning, structuring and schema generation.

Parameterization, prototyping, and optimization are just as valid in engineering design decisions as they are in quality control premises. Both designing and quality control depend on:

- The retrieval of tolerances and designs from the engineering database,
- Intelligent database access, with agile interfaces including icons,
- The analysis of specifications based on functions rather than shape, and
- Historical qualitative and quantitative references that should be considered.

Figure 13-5 exemplifies this procedure which goes from identification of requirements to quality assurance at the factory floor and equipment reliability in field operations. The premise is that quality cannot be controlled in the system; it has to be built in the early design stage or it will not be there.

With this premise in mind, AI constructs can be instrumental in three ways: They can prompt the designer to embed quality into his product; they can act as auditors and fault finders during design reviews; they can help to establish and observe a valid quality assurance policy.

5. EXPERT ASSISTANCE IN DESIGNING AND THE QUALITY ASSURANCE CHALLENGE

Let's start by taking notice that every step in the design process has its share of vagueness and uncertainty. Product designers configure objects under constraints, comparing tolerances to product vulnerabilities and their associated costs. These tolerances will eventually serve for diagnostic reasons, implying system malfunctions and prescribing remedies.

Prediction will be done on the basis of quality variety under given situations and control will be exercised by interpreting, predicting and monitoring behavior. This process will be less reliable if:

- Many valid alternatives and selections are discarded out of hand rather than experimented with.
- Field feedback is not taken into account prior to design proper or is in a form that cannot be easily integrated.

- Diagnostics and quality assurance methods are not properly embedded in the engineering design concept.
- The designs of products and processes themselves are properly integrated.

All four factors negatively affect product quality as well as the cost of products and processes. They also inhibit the ability to continue to develop them through flexible evolution.

Yet the benefits of quality assurance have been time and again dramatized by the fact that improved quality leads to improved productivity through decreased mistakes, more efficient management, less reworking, happier workers, and satisfied customers. Proper use of statistical methods helps:

- To locate quality problems,
- To identify the roles of managers, professionals and clerks in the quality improvement process, and
- To measure the effectiveness of action taken.

Everyone must participate in the quality improvement effort. Greater emphasis should be placed on management's commitment to change its way of doing business and to focus on product quality from the level of the drafting board onward.

Corrective steps can always be taken during the manufacturing process, but they require much more expertise and such expertise is not widely distributed. This is another way of underlining the wisdom of developing and implementing expert systems, networking them among all workstations whose work impacts on quality as well as quality databases.

This work has to be done in a consistent and dedicated manner in spite of the fact that most expert systems shells available today are oriented towards a new approach to programming but not to engineering design and quality assurance as such. However, some means for integration have started being developed.

SOLO, for instance, is an AI-enriched platform that can effectively support the silicon designer. Its concept is also extended to a number of other design approaches, particularly to provide a *design critique*.

Just the same, DESIGNWORLD aims to bring knowledge engineering to the design of devices with the goal of increasing the robustness of the system:

- It provides a common store of information about a design, which can be used at later stages for manufacturing and maintenance.
- The overall aim is to create an automated engineering system for small-scale electromechanical devices.

A similar role is played by SIGHTPLAN (1989) in regard to power station site planning. This expert system is blackboard-architecture that can project the temporary facilities' layout on a large civil engineering construction site. It has been applied to a project for the construction of the Inter-Mountain coal-fired power station in the United States.

Another expert system, EDS, helps in fuel injector design. FORBIN is used in assisting small factory representation. O-PLAN employs declarative formulas and formalisms to identify a range of project planning tasks:

- It uses a search-based consideration of alternatives, and
- It supports cooperative work between men and machines.

In all projects along this line of reference proper knowledge of the domain is very essential. The same is true about the able use of a graphical layout to assist both the product designer and the quality controller.

The overall aim should be to build an integrated, knowledge-based system able to support both a product and a process life cycle. This concept underpins every approach to designing in manufacturing, fully integrated product design, manufacturing engineering, fabrication proper, quality control and robotics—including configuration, programming and monitoring.

6. THE PROCESS OF ONLINE DIAGNOSIS

The diagnostic process is multiperspective in the sense that the overall picture of the situation is constructed from elements recognized in various subsystems and functional definitions. The diagnostic process is often hierarchical in structure: Low level problems are used as the building blocks of higher level ones.

To a considerable extent, diagnosis is a classification process that obeys taxonomical principles. The latter are among the pillars of knowledge engineering:

- We aim to classify an object, event, or situation into categories that may or may not be mutually exclusive.
- Data are sequential or parallel in structure, sorted and compared through analogical reasoning.
- This is a knowledge-enriched approach based on rules and episodes.

Our goal in using expert systems in a diagnostic process is to substitute knowledge and the judgment of domain experts through artifacts. Subsequently, we aim to enrich the latter with functions that involve vagueness and uncertainty yet obey technical and statistical principles.

AI constructs bring homogeneity and a fair amount of objectivity to the diagnostic process. This contrasts to the action of humans where different people may have different strengths and weaknesses in their quality control skills while overall they are classified at the same level of expertise. For instance, person X may be better in lateral thinking, and hence stronger in generating alternative hypotheses while person Y may be better in analytic vertical thinking, and therefore, better at exploring a given set of alternatives.

Both vertical and lateral thinking approaches can be exploited with the same AI construct, with the use of knowledge engineering aimed at exploiting procedural knowledge about:

- Systems operations,
- Configuration factors,
- Online observations,
- Calibration, and
- Maintenance.

A great deal of this activity is done in connection with failure detection and test generation. The aim is to devise knowledge representation approaches whereby failure events can be analyzed by merging diverse sources of information: analog and digital signals, logical variables and test outcomes, text from verbal reports, or inspection images.

The active use of expert systems can help in easing the operator work load when interfacing with:

- The engine under investigation,
- Test equipment, and
- Software packages for reliability assessment.

The applications domain includes electronics failure detection, control systems testing, analysis of intermittent failures, false alarm reduction, test generation, as well as maintenance training. The automotive industry has been one of the first to use expert systems in such a diagnostics sense.

Nissan Motors and Toyota, for example, have developed fault diagnosis expert systems that can be classified into:

- On-board diagnosis, and
- Off-board diagnosis.

Both the support of the manufacturer's experts and the accumulated knowledge of dealership mechanics have been found indispensable for achieving improved valid diagnostic techniques and greater efficiency.

AI-based off-board diagnostic expert systems are being developed in this context. The recent trend is toward building service support systems based on a network that connects a diagnostic system, dealership mechanics, and the manufacturer's experts.

The on-board diagnostic system by Nissan Motor focuses on the centralized supervision of electronically controlled engines. The operational condition of an engine is detected by various sensors. Fuel injection, ignition timing, and idle rotation are optimally controlled to satisfy the required operating specifications.

General Motors and Ford are also using expert systems for diagnostics. As a GM executive was to suggest, a diagnostic expert system for the automobile industry must be able to add or update knowledge in a timely manner: It must operate in every garage or dealership, but it must be able to access a network to facilitate dealer feedback and to get support in case the problem is beyond the rules embedded in the local expert system.

With the diagnostic AI constructs currently in use by the automobile industry, malfunctions such as a bad start and engine failure can be detected and the conditions of these malfunctions properly established. Simple testing then pinpoints the malfunctioning component and identifies possible failure points.

In the shipbuilding industry, Ishikawajima Harima has applied AI in its total ship management system, which is now being marketed. The expert system consists of:

- DIMOS, an online operation monitor, and
- A preventive maintenance and stock control construct.

If an abnormality occurs during ship operation, DIMOS provides necessary on-ship information and makes it possible to take quick corrective action. This expert system contains a preventive maintenance subsystem that uses AI technology to produce maintenance reports and work schedules based on equipment condition. It works ad hoc rather than through fixed time intervals.

Space research and exploration is another domain of implementation for on-board and off-board expert systems. Ishikawajima-Harima, Mitsubishi Heavy Industries, Nissan Motor and computer manufacturers are participating in the development of AI applications in space-related diagnostic systems.

In Part One, we spoke of ACE, an expert system for preventive maintenance of telephone cables whose action is based on a knowledgebank and a database containing repair activity records. DELTA/CATS is an expert system that troubleshoots for locomotive engines, solving a single type of problem: diagnosis.

MDX is a distributed problem-solving construct based on the notion of a group of specialists organized in a hierarchical structure. Apart from the diagnostic side, MDX includes PATREC and RADEX, interacting with these two auxiliary expert systems to obtain and interpret historical data and laboratory references.

Patented after MDX is AUTOMECH; it diagnoses automobile fuel systems. Developed by IBM, DART is a diagnostic expert system that uses functional models of the components instead of rules to diagnose a problem. This is one of the earlier AI constructs that uses deep functional knowledge.

IN-ATE is a probabilistic electronics troubleshooting expert system following a binary pass/fail decision tree of test points. ARBY/NDS is a diagnostic inference scheme for avionics and communications networks troubleshooting, using a hypothesis refinement algorithm.

LES is a rule-based expert system for electronics maintenance. STAMP is an avionics box failure detection construct with test sequences organized by:

- Failure history, and a
- Dynamic modification of the fault tree.

Developed by DEC, IDT uses CAD to speed up diagnosis. By contrast, APEX employs truth maintenance tables and a mixed chaining approach. If there is a likely goal, it employs backward chaining from it, if there isn't such, it employs forward chaining.

RECONSIDER is a diagnostics construction tool. RAFFLES diagnoses computer faults during systematic maintenance. Another expert system, SPEAR, analyzes computer faults while the machine is in operation. MIND reports about tester diagnostics with a hierarchical type representation.

Diagnostic and test selection shells have been developed and used for integrated circuit testing, data communications monitoring, and avionics maintenance training. Some use analogical reasoning, others cause-and-effect failure analysis. Inference is often by truth maintenance tables with propagated constraints and a set of domain-independent diagnostic metarules.

7. USING EVIDENTIAL REASONING IN FAILURE DETECTION

Messerschmitt, Boelkow, Blohm (MBB) has chosen *evidential reasoning* as its knowledge-based approach in quality assurance. The use of AI in this domain has been prompted by the increasing complexity of spacecraft as well as very stringent safety requirements.

These goals led to a sophisticated implementation known as Failure Diagnosis, Isolation and Recovery (FDIR) which operates both on board and on ground (off board). After an analysis of the inadequacy of probabilistic methods, a methodology of *plausible reasoning* was developed by MBB/ERNO for the FDIR system of the European spacecraft COLUMBUS.

Evidential reasoning is plausible reasoning and in this particular case the process focused on failure diagnosis, annotation and recovery. "Our increasingly complex automation aggregates require steady vigilance through AI," said the responsible executive. "And this is as true of diagnosis as of recovery." The main requirements are:

1. *Completeness.* Knowing all possible operational states and all anomalies that may occur.

At the same time, however, it is absolutely infeasible to foresee all ocurrences over a life span. Hence, we have the need for learning systems.

Completeness requirements are constrained by system complexity. Basically, they involve both explicit knowledge and implicit knowledge. Examples of the former are anomalies and their expected manifestations. Examples of the latter are failures that could be handled through model-based reasoning.

2. *Stability,* including the full coverage of all possible manifestations of all anomalies.

The stability requirement is fundamental but also constrained by the fact that the real behavior of complex technical systems can never be anticipated exactly. Examples are locally erroneous data due to: Sensor degradation, errors in monitoring, or errors in data transmission.

Other factors working against stability are the weak manifestations of anomalies, as not all symptoms are clearly evoked. Also the transient evolution of anomalies and the predefined symptom pattern usually become apparent only in the final stage.

Real behavior wrongly preconceived (even in a local sense) and multiple failures with locally counteraffected sensors create *brittleness*. This results in system failures or interruptions due to small deviations, deviations being events that should not happen. Hence, there is significant interest in using knowledge engineering for the able identification of anomalies that can be:

- *Explicit,* with everything stated in advance, or
- *Implicit,* to be handled at a metalevel, thus permitting knowledge-based reasoning.

For this purpose, the implicit characteristics and constraints have to be mapped into a model, for instance, the generation of patterns. But also it must be appreciated that anomalies cannot and will not manifest thmselves as preconceived. "Some of the symptoms we have thought about will never occur—but others we did not account for will come up," suggested the cognizant MBB executive.

All this has important methodological implications. A knowledge-based approach is required which intelligently handles:

- Incomplete or locally erroneous knowledge,

- Incomplete, vague or locally erroneous information (hence uncertainty); as well as
- Multiple failures and their patterns.

Essentially what is needed is an appropriate method of plausible reasoning based on given evidence. This is where AI comes into play.

It is precisely this situation which led MBB/ENO to the adoption of a method of discernment-oriented reasoning based on the notions of discrimination and similarity. Use has been made of both symbolic and numerical representations, where the latter does not require the entry of subjective certainty factors but is based on specified symptom patterns. Such a process facilitates the task of knowledge acquisition.

8. ASSURING THE SUCCESS OF KNOWLEDGE ENGINEERING IN DIAGNOSIS

The applications example with MBB/ERNO, as well as the other references given throughout this chapter, help document that failure detection, testing and maintenance are knowledge-intensive tasks. While test procedures and maintenance manuals contain recommended detection, localization, testing, and repair actions, their use alone does not guarantee the successful completion of diagnostic chores:

- Apart from employing test procedures and maintenance manuals, skilled troubleshooters use heuristics.
- They also have a good understanding of how the system works.
- It is this *metaprocedures* type of knowledge that enables them to perform at an exceptional level.

The top diagnostics and maintenance technicians have a significant familiarity with procedures but also an understanding of symptoms' interactions as well as of the relationships between symptoms and failed components. They know how a given product works and exhibit an intuitive appreciation of how it will behave under certain conditions, including failure.

This sort of expertise expressed in the rules and patterns followed by the human specialist in quality assurance, diagnostics and maintenance is mapped into the AI construct. As a result, when confronted with a problem, the expert system analyzes it in a structured manner, though it may use heuristics to explain all possible alternatives.

In principle, expert diagnosticians have their knowledge organized in powerful knowledge repositories that enables them to reason quickly from given symptoms to specific system level problems. To do so, they use various testing procedures—and this is precisely what the expert system is doing.

The problem of representation of diagnostic and maintenance knowledge is closely related to how this knowledge is used to solve quality assurance problems:

- Determining new facts based upon what is already known.

- Assuring a quick and easy retrieval for adequately expressing similarities and differences.
- Reasoning about solving a specific problem, by looking through its different aspects.

To satisfy the various requirements for the knowledge representation, different techniques have to be used. Selecting the appropriate knowledge-handling methodology is important because a problem-solving strategy greatly depends upon the representation scheme.

As the MBB/ERNO case study documented, basic requirements for failure detection as well as testing and maintenance are:

- A minimum nondetection probability,
- The avoidance of false alarms,
- An interactive usage mode, and
- The ability to integrate knowledge.

This allows for the effective use of procedural heuristic knowledge as well as the exploitation of historical quality data and oncoming information.

Since the art of knowledge-based systems for quality assurance is at its beginning, the types of problems presently solved are in a more or less narrowly defined area of expertise. Within this perspective, the expert system:

- Is capable of exploiting inherent redundancies in machine behavior,
- Accounts for limitations of a particular maintenance procedure in terms of fault isolation, and
- Recommends repair actions and schedules them by reflecting on personnel and test equipment availability, skill level, available spare parts and so on.

I have as well underlined the need for a *quality database* that should be easily expandable to accommodate new references, incoming data, and changes in current content. This quality data base can be nicely explored by using metaknowledge.*

The knowledgebank associated with the diagnostic and maintenance expert system must be capable of handling uncertainty. In quality assurance uncertainty comes from many sources as was discussed in section 6. Just as vital is the generation of explanations and ad hoc query capability.

For the diagnostic expert system to be usable by the maintenance personnel, who typically possess different levels of skill, a human window with sophisticated facilities is very essential. The expert system must be capable of:

- Explaining its line of reasoning,
- Justifying the advice it gives to the user,
- Allowing the user to query its knowledge bank to gain a better understanding of the maintenance task.

---

* See also D.N. Chorafas, *Knowledge Engineering*, Van Nostrand Reinhold, New York, 1990.

Finally, the diagnostics and maintenance AI construct must have a highly efficient control structure. Its purpose is to determine priorities and to decide which subtask should be done next. The issues of control and communication are interrelated. Global control determines when to start a test and diagnosis process, as well as when this process is complete.

14

Selling the Products That We Have Made

1. INTRODUCTION

Marketing evolved as a result of applying management concepts and research initiatives to distribution and selling. Precisely, this involved *selling* the products we made to markets and engaging in competition with other firms that make similar and sometimes superior products.

The *quality* we build into our products is the pride of our work. It is also our competitive advantage against the competitors we encounter in the marketplace. The product is for the consumer, and so is the quality embedded in it.

A desk is a dangerous place from which to view the market and study its behavior. Salesmanship means *movement,* and this is true all the way from the study of market trends, to the analysis of our competitors' moves and products, the examination of sales practices and after-sales service policies.

Selling is not just done through headquarters. Our sales offices and our dealers are the outlets in contact with the market and through which the sales effort is made. For many products, customers need hand-holding not a one-shot sales pitch.

If our salesmen cannot sell the products we make, then our company is in trouble. Or the problem may lie in our network of dealers. They may, for instance, find the complexity of our products daunting and tend not to deal with the kind of customer we are aiming at.

Compaq seems to be in the middle of such a problem and as a *Business Week* article was to suggest:* "That spells trouble for a company that since its 1982 founding has relied solely on resellers. By contrast, IBM and most personal-computer makers split sales (of personal computers) between a direct sales force and retailers."

---

* July 2, 1990.

Compaq's dealers are accustomed to selling $3,000-to-$5,000 machines. But the computer has grown over the years, and when fully loaded with items including extra memory, chips, and storage space, the Systempro can carry a price tag of as much as $20,000 and can require extensive training and support from the dealers.

Such a beefed-up product also calls for a different marketing approach: Selling to higher-level corporate customers than retailers are familiar with. Can technology be used to aid the sales function?

The answer is that it can do so quite effectively. Expert systems have been instrumental in marketing for a number of reasons. They:

1. Assist the salesman to develop the proposal to the customer gaining points over competition, as we saw in chapter 6 with DEC's XSEL.
2. Provide an enabling computer-to-computer linkage between sales and production planning, leading to computer-integrated manufacturing, as we will see in chapter 15.
3. Help in product quality improvement through field feedback, reliability studies, analysis of failures, and so on—a knowledge-based approach to the study of field reports.
4. Assure a number of supporting functions, some of them being major contributors to good salesmanship.

For instance, expert systems have been successfully used as customer profile analyzers, market analyzers, intelligent charting instruments and the like.

All these contributions are particularly important to companies whose management is not content to get its information through the corporate chain of command via interoffice memos but wants to get to the heart of the matter. That is right down to the details of the partnership connection with the customers, through effective hand-holding, keeping every account under steady review.

2. MARKETING STRATEGY AND CORPORATE OBJECTIVES

In marketing and sales, as in every other implementation domain, AI is not a goal but a tool. Like any tool, it can serve best if we first sort out our objectives. When Jack Tramiel got hold of Atari, he articulated the basic principles to guide operations in the following way:

- To offer the latest technology at an affordable price.
- To sell to and buy from the world, taking a global view of the market.
- To sell to the masses, not the classes.
- To believe in a fair profit for the firm, its suppliers and its dealers.
- To maintain a horizontal management style and a lean overhead.

Tramiel's underlying philosophy has been that business is war. It is marketers who carry the word "campaign" from battlefield lingo to the consumer's vocabulary. And as in war, the officers and the soldiers should have superior weapons. That is where AI comes in.

TABLE 14-1 Job Description of the Director of Marketing at the ABC Manufacturing Company*

Responsibilities	Indicators	Objectives
1. Develop a new marketing organization	Establish exact objectives by area of operations	Within "x" weeks from start
2. Hire able persons	* Set up a regional office	* in "y" weeks
	* Set up 10 provincial offices	* in "z" weeks
3. Open market perspectives	* Communicate with government agencies	* within "x" weeks
	* Penetrate major districts	* within "y" weeks
	* Market by industry	* within "z" weeks
	* Enhance public relations	* within "w" weeks
4. Get new customers	* Present imaginative contracts	Establish quotas:
	* Show sales results through letters of intent	* how much
		* how fast
		* which product(s)

* Established yearly. Reviewed every 3 months.

In marketing as everywhere else, expert systems are not made in the abstract. They are written to support strategies and management policies that must first be spelled out. For this reason, in the present section we will stress corporate objectives, and when this message is properly understood we will examine how technology can bring these objectives forward.

Product innovations as well as forward leaps in sales practices come from probing research. "We want to know how the car fits into people's lives," says Steve Barnett, director of product and market strategy for Nissan North America.

Barnett, a PhD in anthropology, hires teams that go into US homes with video cameras, taping interviews and surroundings down to kitchen cabinets, sometimes paying people $50 to cooperate. Then, Nissan researchers translate the findings into subtle new features.*

Such probing by market-oriented companies does not stop after the car is out. Each Lexus employee in America must call at least one Lexus owner each week. Key to such close customer contact is what Mazda chairman Yamamoto calls *kansei engineering*, which he defines as absolute awareness of both reason and emotion.

To provoke engineers' imaginations, Mazda runs consumer focus groups where people talk about everything from food to fashion. Other companies follow similar policies with their salesmen, many involving quality circles to help in sensitivity analysis.

Self-respecting companies also establish properly tuned job descriptions that are dynamic all the time, an example being given in Table 14-1. Expert systems can be successfully used to follow up on the marketing and sales responsibilities defined by such job descriptions, presenting a plan/actual evaluation on demand.

Another area knowledge engineering can be instrumental in is the steady evaluation of the return on financial allocations by *our* firm and by its competitors. There are reasons to believe that the more a company spends on marketing, the more there is left for profits—if

* *Business Week*, October 22, 1990.

the production expenses are very carefully watched. This is shown in Figure 14-1, and it is a fundamental choice of management.

Manufacturing companies are often classified as market driven and product driven. The fact remains that more emphasis on marketing most often means higher profits. Hence:

1. Technology firms should spend a good share of their income on well-directed marketing activities.
2. The trend line is that those companies that spend more on marketing, spend less on production, and vice versa—as shown in Figure 14-2.

All this is part and parcel of corporate strategy. Louis Sorrell, my professor at the Graduate School of Business, at UCLA, used to say that to learn the strategy of a company we have to read between the lines. That is where the message is.

Following this advice and making pertinent observations of Atari's management style, we can deduce the following elements of the company's operating philosophy: First, *price,* is the most important competitive weapon.

In order to use it in an able manner, Atari aims to be the lowest cost producer and

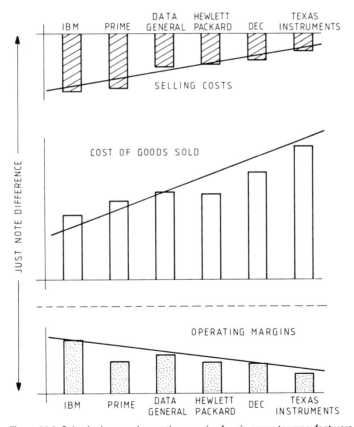

Figure 14-1. Sales budgets and operating margins for six computer manufacturers.

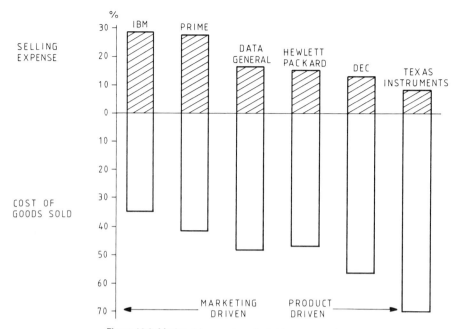

Figure 14-2. Market driven and product driven companies.

distributor of computer products. Its strategy is that if it has the lowest price it should be able to offset the advantages competitors may have due to such elements as added features, distribution network and home-office support.

Second, the world market emphasis Jack Tramiel correctly underlines, suggests that as CEO:

- He will not maximize profit margins, but will
- Try to maximize profits through volume.

High profit margins only create a price umbrella for competition to thrive. Instead, Tramiel said, he will use rising operating profits to subsidize new products that will allow him to become his own biggest competitor.

Other forward-looking chief executive officers also know that innovation and very well-tuned financing play major roles in sales and in marketing. This has led some of them to the definition of a family of expert systems covering a large domain of company operations identified through emphasis on cost control—from manufacturing to marketing.

But technology alone will not solve a company's problems and the same is true of any single-minded approach that tries to avoid harsh reality through gimmicks, for instance, mergers and acquisitions.

Financial problems at Unisys forged from Burroughs and Sperry, have reminded the market that putting two weak companies together does not always result in synergy. In a similar manner, expert systems in marketing will not make a company's products an instant success, but the lack of them may make them a long-term failure.

The message is that the solution lies in the proverbial long hard look, strategies worked out the Tramiel way or for that matter according to the Japanese manner of doing business. Take Philips as the opposite example.

The margins on a traditional minicomputer of the kind Philips used to market have been for some years 50 percent or more, enough to support a substantial direct sales force selling to major users and providing considerable backup and support. But competition saw to it that such margins have dropped to 30 percent or less.

Companies that stayed too long with the concept of fat margins have been washed away. That's how Philips lost $1 billion in one year. The failure is, above everything else, one of policy.

The trick in the computer industry today is to secure the sales channel, the cheapest way of getting the computer to the customer from the factory. Expert systems can be of major assistance in the implementation of such a policy, but first of all, the policy guidelines must be there.

3. AI IN THE SERVICE OF SOUND POLICIES

The AI constructs we build should reflect and support management's philosophy and not contradict it. A fundamental part of Jack Tramiel's success while at Commodore and at Atari has been his management strategy of implementing policies that:

- Refuse to be irrevocably tied to the past,
- Constantly observes the world around it in order to set guidelines for the future.

Such management is always willing to seize an opportunity when it arises, even if it means a 180-degree turn in direction. This type of strategy often confuses associates as well as competitors and financial analysts. But the man at the helm knows that this is undertaken with the goal of achieving long-term success.

As the opportunity arises, companies with a strong balance sheet and a dynamic business philosophy take control of the market and other companies. And by having a worldwide view, they benefit from foreign exchange gains reflecting, as circumstances change, a weakening or a strengthening US dollar.

While some of these ideas may represent a radical departure from the way the majority of corporate America operates and from what most business schools teach, they have a great deal of validity. Therefore, they have been, and will continue to be, instrumental to the success of aggressive marketing companies.

Able marketing starts at the R+D laboratory, and expert systems written to expedite product introduction are as vital in sales as they are in engineering development. New product introductions form an integral part of a dynamic company's operating philosophy. Such a company:

- Continually watches technological developments and assesses how they can be applied in the shortest possible time.
- Steadily plans for cost effectiveness, and does not simply introduce innovative products more frequently.

- Bets on a transnational, well-trained network of sales offices and dealers to channel its products to the market.

Atari, for instance, has about 2,000 dealers around the world, but it also closely watches the grade and rating of these dealers. Different standards are assured for each class.

In the United States, most professional computer dealers are rated A, B or C based on *size* and *sophistication*:

- The "A" dealers are those who service a lot of commercial customers on a continuing basis.
- The "C" dealers are smaller, often selling items other than computers (in fact, in many cases, computers are only a sideline for them), and mostly depending on walk-in traffic.

Products have to match the dealers' strengths and weaknesses and such coordination cannot be made by hand or through classical DP. Expert systems instrumental in exploring information in the database as sales results come in, follow up class by class and dealer by dealer.

Coupled with simulators, expert systems produce sales projections and also guide the hand of the marketing department in directing the dealer's network. When we plan for successful sales we must appreciate that market share will not come as a matter of course and the same is true of profit margins. Without a clear vision and the appropriate tools, we are condemned to fall behind our competitors.

This is as true of the financial markets and the information providers as it is of computers and automobiles. Take, as an example, the Credit Clearing House (CCH), a product group within the Dun & Bradstreet Business Credit Services:

- CCH is dedicated to serving the risk management needs of apparel industry manufacturers, wholesalers, jobbers and marketers.
- It serves this market of about 4,000 customers by assigning credit ratings and giving dollar-specific credit recommendations to their retail customers.

Prior to the use of expert systems, preparing CCH credit recommendations required a staff of trained analysts to review data from several different D & B business information reports to maintain and update a database of credit ratings of approximately 200,000 businesses.

The quality of this specialized service for apparel industry customers depended not only on the staff's analytical skills, but on assuring that recommendations reflected the latest data base changes. Another key element to market success is the ability to respond quickly to customer queries.

To assist in the performance of these marketing functions, an expert system was developed on a personal computer and then ported to two supermicro which act as co-processors to the data base engine. The AI construct accesses realtime information stored in CCH's data base to provide dollar specific recommendations to its customers within seconds.

By dynamically updating available information, the expert system improves CCH's accuracy and response time in handling requests. As a result, its customers can be confident in establishing credit lines for the retail stores with which they do business.

Prior to implementation, the knowledge engineers developed a prototype of the expert system and demonstrated it to senior management. Conclusions established that:

- Many components of the CCH credit analysis were generic and could be reused for a variety of products; and
- The time needed to develop system requirements for analytical products could be significantly reduced through a knowledge-modeling approach.

To make this expert system as comprehensive as possible, over 100 test cases reflecting a wide array of different circumstances were entered at the prototype level. Once all the pieces were in place, the project team began a volume test of 2,800 cases in the CCH database—approximately 2 percent of the total number.

An initial review of the volume test results showed a 92 percent agreement rate of the experts with the response given by the AI construct. After fine-tuning, the rate was raised to 98.5 percent. In addition, certain *knockout* rules were defined. (A knockout is a case for which a systems decision cannot be accepted as completely accurate, and it is referred to an analyst for review.)

Two types of decisions are made in the CCH application:

- One type offers a recommendation and a dollar guideline.
- Another is a no-guideline decision which results when the expert system determines that it is unable to offer a valid guideline on the basis of available information.

When a case is determined to be a knockout, different decisions are produced by the expert system based on the analysis that has been completed. This represents an added value to what was done prior to the use of AI, and it is characteristic of what we get through expert systems in marketing.

4. TECHNOLOGY AND THE STORY OF MARKETING

Marketing is an aspect of business that touches all of us. Few can escape the assault of commercials, promotions, new packages, and shelf displays that manufacturers and retailers depend on to attract buyers. The evolution and fine tuning of such techniques as distribution and the building of brand names distinguishes marketers, sorting out the successful ones from the unsuccessful ones.

If we look back in time and follow the more ingenious sales efforts, we could divide the history of marketing into six phases, starting with the late nineteenth century:

1. At first, marketing was restricted to order taking.
2. Then it started to emphasize a local sales pitch.
3. The advent of railroads and motor vehicles enabled manufacturers to ship goods efficiently and inexpensively throughout the country.

This led to the creation of the first national *marketing campaigns* as well as to the concretization of marketing concepts necessary to face intensified competition. Literally, intensified competition meant that manufacturers entered into each other's domains. As a result, during the second and third quarters of the twentieth century:

4. Techniques were developed to divide consumers into *demographic* and *psychographic* groups, further intensifying competition.
5. *Global marketing*, the next phase, came during the last 30 years, and it capitalized not only on the media but also on the wave of prosperity that swept the First World after World War II.

How smoothly executives made the transition from one phase to the next has been the cornerstone of business success as the marketing battles fought by Coca-Cola and many other vendors demonstrate. Coke was famous for its ubiquitous advertising, leading other vendors—from the auto industry to the home appliance business—to use advertising jingles.

This has been the strategy of mass marketing, but a new era is now starting which many industries have not yet appreciated and therefore are going to miss:

6. The *personalization* of the product and services done over a wide range of businesses from motor vehicles to banking.

More than ever before in history, in the years to come, sales productivity will depend not on one single issue but on a compound effect. As Figure 14-3 suggests, this rests on three axes of reference: communicating, designing and personalizing. Heuristics, computers and networks can be instrumental in promoting every one of them—and most particularly, the personalization perspective.

A good example on the role expert systems can play in the marketing of financial services is offered by Citibank's Quotron. Operating on line:

- An AI construct acts as a profile analyzer.
- It learns the likes of the individual treasurer, and
- It follows a presentation of financial data according to the treasurer's viewpoint rather than the general good-for-all but average approach.

Today, personalization is the key word in competition. Even the soundest marketing strategies can be undermined by managers who refuse to change with the times. Henry Ford's obstinacy about market segmentation, for instance, is legendary.

The father of mass marketing refused to believe that the Model T had ceased to fit all car buyers' needs. This blind spot gave General Motors the advantage when Alfred P. Sloan, Jr. began targeting specific models to different types of consumers.

Slowness in adapting to changing market tastes and drives has created hardships for other firms as well. Asa Griggs Candler, the first chairman of Coca-Cola, resisted selling the soft drink in bottles, seeing no need to expand beyond the company's fountain business. It was Candler, however, who instituted the company's tradition of monstrous advertising budgets to increase consumption, which set the pace for other companies.

Over the years, this practice has boomed but also the realization set in that advertising budgets can be used in more intelligent ways than by using brute force:

- Toyota has installed in its Tokyo salesrooms networked CAD stations that permit the customer to design his own auto on the screen.

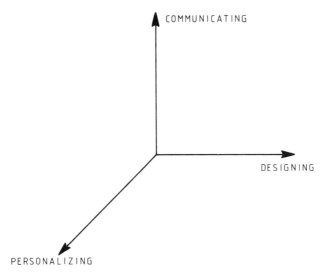

Figure 14-3. Axes of reference on which depends sales productivity.

- In its Munich showrooms, Fiat uses optical disks to show the customer how its car drives on icy roads and in the desert.
- In its small shops that cannot physically carry a large number of items but can display them on a screen, Compuserve/Compucard installs interactive videodisk units enriched with expert systems.
- The German Bundespost Telekom (telco) effectively uses expert systems in its sales offices to help the salesgirl channel telephone products to the customer and prepare networking proposals without having the technical background to do so.

The able use of AI can ease the pressure on busy executives who want to do a great deal of work by themselves. Richard W. Sears, for instance, spent so much time writing the copy for his mail-order company's catalogs that he ignored the need for orderly management techniques to accommodate Sears's stupendous growth.

In many cases, management comes to expect massive internal changes at its own company. But:

- Looking over your own shoulder is not enough.
- Looking over your competitor's shoulder is not enough.
- If you are going to be competitive, your salespeople and CEO must look over their customers' shoulders as well.

Expert systems can make a significant contribution in this direction by analyzing results and keeping the focus near sales.

Most importantly, high technology can assist those marketers who learned to create a need, persuading companies and consumers not only to change their buying patterns but also their most deeply embedded habits. Equipped with networked portable computers whose software is enriched with expert systems, salesmen can present the customers with alternatives and with details that cannot be done through paper or words.

Expert systems are an able answer to the query: "How will today's companies adapt to an even more segmented market that has become harder to reach through traditional media?" As the use of traditional media becomes more widespread, they lose their competitive advantage. We must always work to improve what we did yesterday.

High technology is the way to utilize effectively the reams of consumer information now available to manufacturers—from market research to scanner-type cash registers. Business now needs innovators for marketing's next phase, and this, as we said, is *personalization*.

5. SALES, DISTRIBUTION AND FIELD SUPPORT

Expert systems encapsulate the knowledge of problem solving associated with a given product and its sales. The help provides advice on problem identification, the counting and monitoring of client responses, as well as the minimization of a future effort to close a sale.

The concept behind sales assistance through AI modules is remarkably simple. The expert systems designer can take a piece of text, such as a complex set of instructions or regulations, and:

- Separate the actual sales advice from
- The conditions that define when that advice is applicable.

The expert system will not only display options but will suggest the ways and means for proceeding with the sale as such.

There is, however, more scope in AI implementation than in the sales proper and this has to do with the overall organization of design. More specifically, this involves how we structure our efforts so as to respond to market drives while we reduce inventory and transportation costs and at the same time increase our ability to meet demand.

About one in five American suppliers now receives its orders from customers directly through a computer. American Hospital Supply, a company specializing in medical equipment, has created particularly strong links with its customers by providing them with computer terminals on which they order all their medical supplies. To switch to a rival, American Hospital's customers would have to bother familiarizing themselves with a different system.

This type of vendor-customer relation can be effectively strengthened through knowledge engineering. The higher the level of technology used, the more difficult switching becomes—provided that the organizational perspectives were put right in the first place.

Organizational design in the marketing field starts with the analytical study of:

- A site location for sales offices,
- Distribution policies, and
- Field support centers.

It aims at establishing goals and in following up on goal deviations, going all the way to telemarketing but also emphasizing the mechanics of the sales function:

- The acquisition of sales orders,
- Their verification for acceptance,
- The fine tuning of pricing a sales offer,
- The perspectives in inventory control,
- Transportation planning and dispatching, and
- The optimization of expediting routines.

An expert system assists DEC in reducing transportation costs and it also improves truck planning and delivery. This AI construct uses a rule-based representation to map the know-how of the dispatching people.

Part and parcel of this solution is the continual mileage planning and scheduling. The input to the expert system consists of transportation requests and transportation contracts. The output is transportation assignments. Steady use saves 10 percent of the transportation costs.

Along similar lines, another expert system has been designed for *truck deliveries* and it has been implemented by a number of firms. This model:

- Computes distance,
- Evaluates loading capacities,
- Estimates time constraints,
- Considers the quality of roads, and
- Integrates the maximum allowed speed as well as other factors affecting delivery.

It supports color graphics with an on-screen presentation of shapes, tracking routes, and stops by trucks, as well as an economic evaluation of alternatives.

Among the users of this construct are Air Products and Chemicals (liquid oxygen, 3,500 customers, 22 million miles); Exxon; Hershey Foods; Edward Don (a restaurant supply with 1,100 customers, which saw an efficiency increase of 10 percent); and Dupont's Clinical Systems Division. The latter had 1,500 customers in 1,000 cities in the US and Canada; with the model, it now serves 3,000 customers.

Pinpointing where customers lie on a distribution network helps suppliers decide what action they should take to boost or to retain sales. Expert systems can help to:

- Identify the things that impress particular customers such as price, the speed of delivery, the quality of service or a technological lead.
- Evaluate what costs and risks a customer would incur if he or she were to switch to a new seller, in search of better cost-effectiveness.

The higher the cost and the bigger the risk in switching, the less likely a customer is to desert the supplier. In some instances, clever vendor firms bind their customers to them by hooking them into their own ordering, delivery or inventory systems.

In Chapter 13 we saw how AI technology can significantly assist quality assurance. This has its obvious consequences as to the role expert systems can play in field support, as a number of real-life cases help document.

Today, IBM maintenance engineers are equipped with portable computers equipped with expert systems that provide significant assistance in troubleshooting. Other applications include realtime monitoring and the diagnosis of steam turbines and generators:

- The input is sensor readings,
- The output is malfunction identification and guidance.

We have also spoken of automobile troubleshooting by Ford and General Motors, and such AI models are in production use. As a matter of fact, engineering and manufacturing diagnosis has been one of the most successful AI areas and this has obvious effects on sales.

6. MERGING TELECOMMUNICATIONS AND EXPERT SYSTEMS

In 1987, United Parcel Service (UPS) devised a scheme to transform its paper-and-pencil operation into an on-line package delivery system. Drivers would no longer call in from phone booths. Instead, 55,000 UPS trucks would be fitted with radio receivers and transmitters, enabling the package carrier to give customers up-to-the-second information about deliveries.

There is also the example of Mrs. Fields Cookies, which in 1980, sold $130 million worth of chocolate-chip cookies in shops around the world—a dozen years after Fields founded the first cookie store. This is a striking example of what online communications can do to sales.

Fields oversees more than 400 stores from Park City, a ski resort in Utah. Over the past few years she has installed an elaborate automated system to manage the stores. The idea is that a solution should take over many of the administrative tasks, leaving store managers free to concentrate on people-oriented jobs such as selling and creating a pleasant environment in the store. This realtime approach:

- Monitors hourly sales,
- Plans hourly production,
- Handles accounting and stocks, and
- Administers personnel tests for hiring.

Through voice mail and electronic mail, store managers are also encouraged to talk directly to Mrs. Fields and she responds to them.

In a way, whether the realtime system is controlling or enabling depends on each store manager's personality. Those that enjoyed the administrative part of their work presumably felt overtaken by the solution. By contrast, those who preferred the people side, felt relieved by the online approach. The principle is applicable to many other industries and further gains can be achieved by betting on AI.

These further gains have been exemplified in several chapters, particularly when we spoke of just-in-time inventories. But it has also been stressed that for industries, zero inventories require not only organization, algorithms, heuristics and computers, but they also need networks.

Financial institutions involved in a worldwide trading environment appreciate that instantaneous communication is vital. As manufacturing companies try harder than ever to push into foreign markets, they also race to expand their private networks. No wonder that many of them are evolving into global systems that virtually elminate the boundaries of time and distance.

The same is true of manufacturing companies. On Long Island, the Honda dealership can order new inventory using a computer to poll the stocking systems in ten regional warehouses in America. These parts centers in turn are tied to Honda's US headquarters in Gardena, CA, and the parent Honda in Tokyo. Hewlett-Packard's global network permits its executives to conduct meetings over a video-conferencing facility.

Both videoconferencing and the use of expert systems help merchandizers gain a sense of direction, as well as hold just-in-time inventories. The guidelines are known:

- Do not buy too much merchandize that has to be drastically marked down later.
- Do not buy too far in advance. In that way, if consumer tastes change, you can react quickly.
- If something is not moving off the floor fast enough, mark it down to make room for fresh goods.

The challenge is how to enforce such guidelines properly so that inventory levels are not too high and if they get out of control they can start to decline immediately. A number of planned sales promotions have this objective in mind, but sales promotions cut prices while simulators and expert systems can cut inventories without unwanted price effects.

By the same token, the choice of the proper distribution policies and channels profoundly affects:

- The nature and cost of a company's sales organization,
- Its selling prices,
- Its gross margin structure, and
- Its commitment to physical distribution facilities.

These in turn will affect production and supply costs. The pricing of the products is a matter highly influenced by competition, and the market leader sets the pace. In the US, for example, loans for cars are now controlled in their pricing by the General Motors Acceptance Corporation.

Like risk evaluation, product pricing is under both the objective and the subjective judgment of responsible officers, and many company managers are starting to realize that better profits cannot be achieved by raising prices. There is simply too much competition in the domestic and global markets. As a result, they act to boost profits by selling more but also, if not mainly, by cutting costs, reducing inventories and improving productivity.

This means that solutions have to be provided that are as qualitative as they are quantitative and as heuristic as they are algorithmic. Stock-outs, excess delivery time, or the excess variability of delivery time all result in lost sales, but none of them can be effectively controlled through classical DP.

Any change in the distribution system influences the basic elements of customer service, and the gains or losses for the company. These effects, while difficult to measure, must be considered part of the real landscape of sales and distribution; hence, AI can play an important role in this regard.

In Section 4, Figure 14-3 brought into perspective the three axes of reference on which sales results depend. *Communicating* means:

- Directing the sales forces,
- Contacting the clients,
- Taking sales orders,
- Answering status queries,
- Giving technical responses, and also
- Billing and associated file handling.

An integral part of the communicating function involves providing pricing information, product information and other references in response to customer queries. These are activities that can be very well executed online, some of them being handled through expert systems to cut touch labor.

Designing means contributions to product development through market research, but also the management of market segmentation. Part of the latter involves developing leads for the sales effort and assisting in market penetration. We have already spoken about how much this goal can be served through:

- Customer profiling,
- Account planning, and
- Resource scheduling.

We have also paid significant attention to the need for *personalization*. To answer its clients' ad hoc queries, General Telephone and Electronics (GTE) developed the Intelligent Database Assistant (IDA) and immediately put it into implementation (the CALIDA project). CALIDA permits GTE executives to overcome the handicap of heterogeneous data bases giving fast responses to ad hoc customer queries—a definite competitive advantage.*

Heuristics, networks and computers can be instrumental in communicating as well as in forecasting, planning, directing and controlling market activities, and in performing factual and documented database searches, credit analysis, pricing proposals, and the statistical evaluation of sales results.

7. PRACTICAL AI APPLICATIONS TO ASSIST THE MARKETING EFFORT

Philips has written an expert system to help its sales people make tenders to customers. Its output is a graphical design and a bill of materials. Its component parts include two types of interfaces:

---

* For more details on IDA see D.N. Chorafas, *Risk Management in Financial Institutions,* Butterworths, London and Boston, 1990.

- For normal users, which essentially means the salesmen.
- For privileged users, or the system administrator and his people.

Rule language has been specifically created for designing communications equipment. Privileged users can add and subtract rules to the knowledge bank and components to the database. The construct's structure is shown in Figure 14-4.

Design started in response to the query "What do we consider the fundamental problems in sales support to be?" The answer has been that this is a dual problem:

1. One side involves order handling and concerns the daily business including the signing of customers' orders, the expediting of goods, and the maintaining of inventory records and billing.
2. The other affects sales forecasting, and the effort to assure better forecasts of item types to help in exploring more efficient production planning opportunities.

Basic to this requirement is an appropriate sales classification that should reflect the nature and structure of the forecasts to be made by the marketing department. As many companies know by experience, there can be found areas of potential improvement directly related to the use of expert systems.

Many manufacturing companies today have come to the conclusion that developments or solutions capable of handling all of the information processing requirements of the firm:

- From planning, controlling and analyzing
- To the operation of selling, manufacturing, warehousing, and shipping

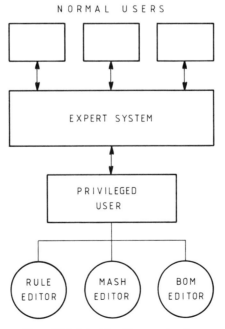

Figure 14-4. A dual function expert system.

have to go well beyond classical DP. As a result, a prominent place has to be given to knowledge engineering technology.

Simulators, too, can be instrumental in fixing maxima and minima based on the monthly turnover of the last 5 to 7 business years, estimating the average seasonally corrected monthly turnover, and calculating the confidence intervals per month.

But whether we wish to use simulators or heuristics, all of the pertinent data presently scattered, and, to a large extent, duplicated among various marketing operations, have to be consolidated. Such activity involves corporate databases as well as an expensive array of interdepartmental communications. Any half-baked solution actually makes decision matters worse.

Another prerequisite is the handling of ad hoc interactive reports:

- Reflecting an alteration of programs,
- Keeping a close watch of individual prices and delivery dates,
- Providing a detailed picture of our company's position in the market, and
- Tackling issues from several different directions in order to achieve polyvalent results.

Quite often, monthly schedules must be broken down into daily requirements so that production orders and material requisitions for the day's operations are available on a timely basis. In other cases, weekly projections are the best approach, and the timing base can vary from one operation to another.

A similar statement is valid in terms of helping the salesman directly in his market effort, as the following case of an expert system written to help in car sales helps demonstrate. From CEOs to salesmen, marketing-oriented people have to realize that their counterparts on the customer side of the table are analyzing their vendor relations a lot more closely these days than they did ever before.

The goal of the AI construct reviewed in the following paragraphs is the selection of a car model with extra features. The expert system proposes to the customer three variants:

- Model choice, which can be paid by the prospect without the latter being a computer expert,
- Model choice, which fits within broader financial limits established by the client, and
- Model and features choices tuned to promote some exciting extras.

The expert system prompts the salesman with queries. The analysis necessary to document the three variants is fed by the responses being received which establish the *client's profile*. This is a rule-based module with the principal queries (and background rules) reflected in Table 14-2.

Figure 14-5 presents in a block diagram the main steps followed by the modules of the expert system. One of the modules analyzes profiles of car models, its tree-type structure being presented in Figure 14-6. Every mode and every leaf has a frame with slots for every key word in the client's profile.

An expert system module estimates the value of a variable outside its tabular or observed range; a small simulator performs extrapolation. An optimizer places the options in the

TABLE 14-2 Client's Profile Analyzer

1. Client's image (old fashioned, modern, open minded)
2. Client's name
3. Client's address (domicile and residence(s))
4. Year of driver's license
5. Client's age
6. Client's sex
7. Client's marital status
8. No. of children
9. Profession
10. Salary or salary group (if loan wanted)
11. Current loans
12. Credit card reference(s)
13. Bank reference(s)
14. Car driven/to be driven by: client, wife, children
15. Any physical disability
16. Miles per year on the car
17. Current car, model
18. Second, third family cars and models
19. Current engine power
20. Wanted engine power
21. Front, rear, 4-W drive
22. No. and type of seats
23. Safety features
24. Environmental issues:
    * exhaust/air pollution
    * noise pollution
25. External color
26. Internal color
27. Type of windows
28. Wanted accessories
29. Suggested accessories (airbag, etc)
30. New car will be used for:
    * transportation to work
    * entertainment and hobbies
31. Travel to foreign countries
32. Export to foreign country
33. Financial data of client:
    * expected price range
    * how much in cash
    * trade-in of old car
    * loan possibility
34. Proposal for salesman
    * budget to propose to client
    * effect of trade-in
    * effect of discounts on profitability
    * impact on salesman's commission

proper or desired order, according to the user's choice. Still another module identifies the instances of difference or the inconsistency.

This application has significantly strengthened the analytical approach taken by the firm in the 1970s in using simulators for the preparation of sales forecasts based on past sales, economic trends, market research, reactions by the local sales managers and marketing management evaluation. Other models address themselves to the coordination of sales forecasts with the company's general inventory policy and the preparation of manufacturing schedules.

The polyvalence of this and similar efforts helps document what technology can do in an enterprisewide sense. The irony is that the more management focuses on cutting, in a one-sided manner, the sales and distribution costs, the less successful it is likely to be in reducing the real costs of sales.

Such an apparent paradox explains why so many companies have diligently pruned distribution costs only to find that these hard-earned savings are somehow not translated into improved profit margins:

- They have been watered down or actually washed out by increases in other costs scattered throughout the company, by contrast,

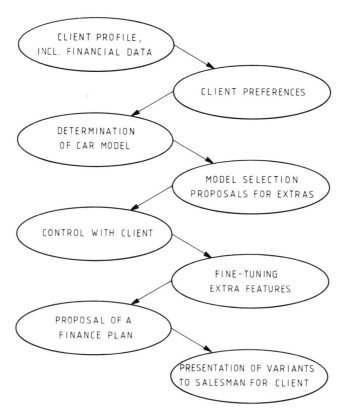

Figure 14-5. Expert system for the support of a car salesman.

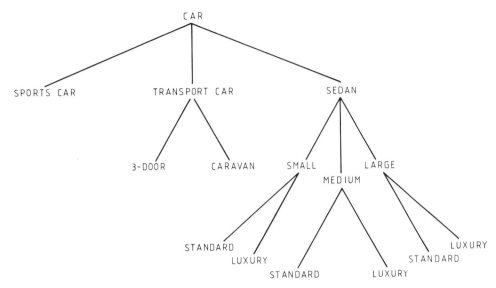

Figure 14-6. Profiles of car models. Menu selection for optical disk.

- A successful integrative effort will range from R+D to manufacturing, sales and transportation, giving all costs a closer analysis, including the issue of their interdependencies.

Though they may never appear as sales costs on any financial or operating report, and they often show up unidentified and unexplained at different times and in assorted places—in development, purchasing, production, and paperwork processing—sales-induced costs often hit anywhere and everywhere in business as a result of mismatches.

When such costs are traced to their roots, we find that they are, in fact, all intimately interrelated, linked together by one common bond. They all result from the way our company sells and distributes its products.

It is this aggregation of the costs, rather than what management considers to be the sales cost proper, that presents an important and increasing drain on earnings. Hence, when we plan for sales performance, promoting the products we have made, we should take the proverbial long hard look across the board, using high technology to reach the market, satisfy its wants, handle the client accounts, swamp costs, and improve profitability.

15

Computer-Integrated Manufacturing

1. INTRODUCTION

The classical structure of a production system is built in multiple management layers. As the outgrowth of classical organization perspectives, such a structure has little to do with computers, and until recently computers have not been used as they should have been employed to eliminate these management layers, flatten the organization, and close the gap between the plant and different islands of business activity.

Computer-integrated manufacturing (CIM) addresses itself specifically to this problem: The aim *is integration in resource management,* establishing a more efficient organization and better cost control perspectives. CIM involves a number of important interdependent areas:

- Production planning and scheduling,
- On-time tracking of production orders,
- Just-in-time inventory control,
- A better use of human resources,
- A reduction of waste,
- An efficient energy administration,
- Quality assurance,
- Maintenance management, and
- A close coordination with sales and distribution.

Some of these issues are long-standing pillars of production management. Others are recent areas of focus where CIM is instrumental, for instance, just-in-time inventories, steady quality assurance and the elimination of waste. Every one of these domains, and CIM, is a fertile ground for the use of expert systems.

Another area where knowledge engineering can make significant contributions in a CIM effort extending from R+D, to manufacturing engineering, production management, inventory control as well as a marketing and sales domain, is *profitability analysis.* As shown in Figure 15-1, this should be done:

- By customer, for all customers,
- By product, for all products, and
- By production center, for each and every center.

But CIM makes no miracles. We need organizational consistency to proceed successfully with this type of application. In other words, the technology is at *our* service but is *our* organization ready for it? Top issues to be faced with CIM are:

- Organizational,
- Cultural,
- Architectural, and
- Technological.

Response planning and coping with the rate of change are underlying issues in practically every COM effort.

When it comes to the implementation of CIM, account should also be taken of the fact that variations do exist by industry as well as within a certain industry. Hence, specialization is needed, though all industries face common requirements: focusing on costs; rethinking and retrofitting the information technology; updating and even restructuring the overall architecture; and integrating the so-far discrete islands of technology implementation into a coherent total system.

To be successful with CIM, not only must we cope, but we must also be somewhat ahead of change in communications and computers technology—even if they grow at 30 percent to 40 percent per year and will keep on accelerating at least until the late 1990s. Therefore, we must focus on *training* and upgrading our human resources. This may sound strange stated in connection to CIM, yet it is the pillar on which able solutions are resting.

2. THE CHALLENGE OF DISCONNECTED AREAS IN MANUFACTURING OPERATIONS

Every company has requirements and adopts approaches for accepting and processing customer orders; it efficiently distributes its products; receives materials; plans production; and holds inventories. The same is true of the execution of financial operations, as well as for R+D.

What many manufacturing organizations lack, however, are the solutions needed for interrelating these various functions, integrating the user environment so that significant efficiency will be supported. The job is *doable* provided we have the know-how to go about it, bringing together disconnected areas and closing the gap that presently exists as shown in Figure 15-2A. This information gap creates:

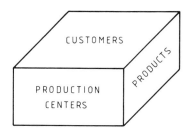

Figure 15-1. A CIM environment requires an integrative approach.

- Delays,
- Errors, and
- Incompatibilities.

Closing it in an effective way is the goal of CIM (Figure 15-2B). As the present and the following chapter will demonstrate, such closing can be effectively done through an enterprise management system implemented in a companywide sense. In the plant itself, it can be achieved through integrative factory information, from networks to databases.

Emphasis must be placed on the user organization and we must appreciate that the users have several functional requirements that utilize, as well as support, potentially integrative activities. Among them are:

- Order entry locations

Sales offices equipped with online terminals can enter or change orders, make inquiries regarding existing orders, customers, and inventory, as well as enter information regarding new customers. For every information element the principle must always be, "One entry, many uses."

- Stocking points

Inventory locations, such as warehouses or central distribution centers may use online workstations to enter shipping information regarding orders executed, scheduled or rescheduled, and to report receipts of stock transfers and customer orders. Freight bills of lading, that is, documents, must be prepared and entered online.

- Planning functions

Headquarters may use queries through online terminals as well as periodic reports to assist in production planning and distribution control, and in monitoring exceptional conditions.

- Accounting

The system should definitely perform accounting functions, providing or receiving accounting data online. Not only should invoices be provided to the accounting organization online, and credit control data received, but this must also be integrated with the general ledger.

As these few examples help document, any successful effort to close the gap between the business system and the factory site must necessarily tackle structural issues both horizontally and vertically. Organizational approaches are typically built in layers that include:

- The presidency,
- Corporate divisions,
- The planning of resources,
- Supervisiory control,
- Machine centers, and
- Instruments.

Vertically, the organization is divided into specialized departments, some of them closed in a water-tight manner. This magnifies the negative effects of the discrete-islands approach.

Furthermore, while at a given point in time the personnel of a company might have been competent, because of the fast-changing environment, such personnel lose their skills. Current projections talk of the need of spending 15 to 20 percent of our budget on training just to keep up with change.

A. PROBLEM: DISCONNECTED AREAS

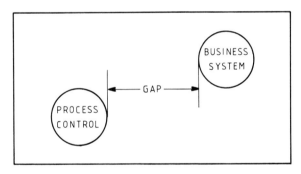

B. SOLUTION: INTEGRATION THROUGH CIM

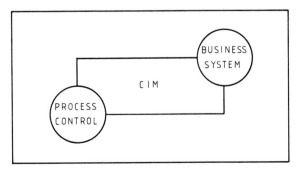

Figure 15-2. Current gaps in subsystem connectivity have to be filled by CIM.

Failure to provide for a steady human retooling will result in our company losing the most important part of its human resources: skill. As a consequence, machines alone, even the most automated ones, will be of no help in:

- Exploiting the existing possibilities for improvements in product quality,
- Bettering productivity on the shop floor,
- Having a truly integrated production landscape,
- Improving management control,
- Swamping investments in inventories,
- Weeding out unnecessary costs, and
- Providing online cost-accounting capabilities.

In most industries, the inability to focus on these problems in an able manner is cultural rather than technical. Yet, organizational rethinking and restructuring is that much more important because of the globalization of business transactions.

The efficient exploitation of global markets calls for internal homogeneity, contrary to the policy of many multinational as well as transnational companies today that solve problems by country rather than by providing solutions for the integrated market to which they appeal. Their management fails to realize that the critical success-factors today are not those that prevailed in the past.

Among the crucial factors to be watched carefully, we distinguish overall improved margins; and a sharply reduced costs; faster billing and receivables; and a shorter time from conception to production. Another vital factor is effectiveness in the transition from production to distribution.

In fact, one of the top benefits we want from CIM is *integrated coordination*. The tough issues we have overcome are those of:

- Getting started in rationalization,
- Providing for further-out cultural change,
- Developing the necessary vision,
- Establishing the basis for cost justification,
- Replacing the antiquated and parochial accounting systems, and
- Staying with a dynamic plan that covers all sectors of business.

The emphasis on new accounting principles and systems should not escape our attention. We need new, more detailed, but also flexible, cost-accounting concepts and procedures, able to integrate and work well with CIM. The old cost-accounting system is based on direct labor and direct materials (DL+DM) which reflects little on the new perspectives that stress *innovation* and *quality*.

As the introductory chapters to this book emphasized, *touch labor* has shrunk, in many cases to about 5 percent of the cost of production. In some plants, labor is fast becoming equally trivial as other minor expense chapters. But this does not mean that labor is "out."

What remains from the labor component is to a large measure the contribution of highly skilled individuals where brain power rather than man power dominate. Without high manual labor costs to justify investments in the more classical mechanization, manufacturers are being forced to rethink their whole approach to profit and loss management. Financial institutions will follow the same path in a few years.

3. A CRITICAL LOOK AT INTEGRATIVE REQUIREMENTS

The manufacturing industry is becoming a loosely structured network of *service enterprises*. Acting as independent business units, these are built around a specialized core of interdependent competence centers which are joined together often temporarily for one undertaking.

This is akin to developing consortia, each with its own network of contract relationships with research, production, and marketing groups around the world but within the same industrial combine. Because of this reason, quite diverse industries are becoming similarly structured, with semi-independent units specializing in design, manufacturing, assembly, distribution, and also system work, networking, databasing and AI support requirements.

With the labor cost chapter shrinking, as we saw in Section 2, the great part of most companies' costs, other than those for purchased materials, typically occur in overhead:

- In manufacturing, more than two-thirds of all nonmaterial costs tend to be indirect or overhead expenses.
- But most overhead is also becoming a sort of service the company begins to buy internally, while retaining the possibility of external supply.

Reconceptualizing the relationship between internal staff and overhead costs as services that could be bought externally exposes a wide gap in *efficiency* which is to a large measure due to overstaffing. This is particularly true of the intermediate layers and is due, to a great extent to the lack of communication—and hence, coordination—among the different staff members.

A lack of coordination is the result of obsolete thinking, but it is also a lack of appreciation by management. When direct labor costs fall below 15 percent of the total product cost, the dynamics of decisions also change:

- Cheap labor is no more a prerequisite for competitiveness.
- The expense of distance, such as freight charges, outweigh the advantages of low wages.
- The skills of the labor force count for much more than easy hire and fire policies, while
- The negative results of a lack of coordination get magnified, propelling the need for integrative solutions.

At the same time, new major investments can be justified not by labor savings but by improved quality, faster product introduction, delivery largely automated, and lower inventories.

Every one of the domains in reference can benefit from AI and this is just as true of human resources management. As an example, the Direct Labor Management System (DLMS) is a knowledgebank for an automotive assembly to be used by managers and plant personnel. It:

- Accesses process descriptions,

- Fills in missing process actions, and
- Constructs entries in a database that contains the precise sequence of human operator motions required for each process.

The expert system uses this information to perform time-and-motion analyses supporting the allocation of labor resources for applications such as factory-line balancing and accurate labor-cost budgeting.

The example is good, but the application must be steadily improved with emphasis placed on polyvalent domains. Today, only backward industries see their direct labor costs as a critical aspect in management decisions:

- In the less organized car manufacturing companies, direct labor costs still account for 25 to 28 percent of the total cost of a car.
- But Toyota and Ford are down to 18 percent and are expected to drop to between 10 and 12 percent by the end of this decade.

Ten years ago it took 6,000 workers to build 1,999 cars a day. Now Chrysler needs about half that many workers to achieve the same output.

- Some electronics companies have already reached the level of 5 percent of labor through robotics, practically standing at par with that of some agricultural sectors.

In other words, technology has provided the foundation to, more or less, weed labor out of the factory, yet companies still do not change their accounting principles. The result is that faulty accounting systems cause errant decisions by management.

Digital Equipment Corporation is a good example of this last reference. Its current accounting system indicates that a certain terminal should be bought rather than made by the company itself, but this suggestion, it has been said, misses the point. Hence, we have the change in policy:

- As internal bids were asked from the factories for submission, a DEC plant manager supplied a bid to make the terminal at his plant and be won.
- The old cost accounting system did not reflect the level of technology and what could be done with it; hence, the same plant manager was a loser.

In another case, a division of Monsanto Chemicals needed to increase significantly the productivity of an aging fiber plant to remain viable. The old cost accounting could not indicate the weakest points but a thorough systems study was able to achieve significant productivity improvements through a combination of:

- Human resources planning,
- Total quality control,
- Just-in-time inventories, and
- Computer-integrated manufacturing.

The lesson is that we must always be looking for competitive advantages and ways to exploit them. A great deal of the challenge involves open communication, precisely what has been underlined by the more recent integrative approaches to manufacturing, properly served through knowledge engineering constructs.

4. KEEPING TRACK OF PROJECT STATUS

Some manufacturing companies are so poorly managed that they cannot account for the cost of many programs and they suffer major breakdowns in their production activities. For instance, in October 1990 a report was made public indicating that the management of major weapons systems by a well-known aircraft manufacturer was so chaotic that:

- The design team working on the engine of the B-2 bomber
- Failed to notify the plane's structural teams of significant changes that could dramatically affect the unusual craft's airframe.

Military authorities also criticized the company for failure to manage relations properly with hundreds of subcontractors and for serious problems in moving weapons from laboratory development to production.*

Although this company is the subject of at least 7 grand jury probes and 11 criminal investigations, the critique of its management and business practices is new. In fact, it is the first time the military service has openly criticized the company in reference which is building some of the air force's most favored weapon systems.

The military criticized the aircraft and weapons system manufacturer as a company that grew so rapidly that its senior executives failed to develop proper controls over its personnel, equipment and ballooning contracts.

Management was said to be rushing into contracts involving sophisticated technology when it had little experience as a prime contractor.

As a result, the company's leadership was overwhelmed with too much too fast, and it provided virtually no training for employees pushed into high management positions. Thus, the report blamed many of the problems on the *culture* of the company.

Among other findings from the air force investigators, this company has been unable to provide accurate cost and schedule estimates for its weapons programs due to a lack of internal controls:

- All the programs reviewed faced major difficulties in transferring from drawing boards and research laboratories to the factory and assembly line.
- There was poor management of the numerous subcontractors; frequently, design changes in programs were not given to subcontractors who provide key components of the systems.
- There were, as well, serious problems in coordinating systems engineering teams, with designers developing different parts of the same weapons system.

While all these ills will not be solved by knowledge engineering, as in this specific case, the salient problem is one of management control; expert systems and networks can make a major contribution after the proper policies are established.

---

* *International Herald Tribune,* October 11, 1990.

Tracking project status is a good example of a domain where significant benefits can be derived through the aforementioned approach: An integrative manufacturing management system addressing the problem in reference will typically focus on:

- Assembly stations, process plans, and jobs,
- The realtime assignment of jobs to stations,
- Production status by job lot, and
- The realtime monitoring and control of manufacturing cells.

Typically, such a solution will operate in realtime enriched by knowledge engineering technology. When we build an integrated system, AI may represent only 10 percent of the effort, but even a little bit of AI can go a long way in planning, coordination and control.

A similar reference is valid in terms of evaluating alternatives. A manufacturing company may have to perform activities at certain critical stages of its value creation process that it would otherwise outsource, if efficiency were the sole criterion. At the same time, as the case of Apple Computers demonstrates, a company need not fabricate many components itself to control manufacturing, though it is very important to:

- Dominate the strategic steps in the manufacturing process,
- Develop internal technological knowledge,
- Sharpen the logistics process, and
- Establish commercial networks able to reach the clients online.

A way to avoid becoming too dependent on critical suppliers is to seek alliances with outstanding, but noncompeting, enterprises. Another way is to develop innovative concepts that can be shared with the business partners at a price.

But putting innovative proposals into practice requires a new culture from managers. For instance, tailoring better systems to manage costs, rather than simply allocating them after the fact, demands:

- A clear vision of a company's long-range strategy,
- A well-rounded understanding of every facet of its business, and
- An intimate knowledge of what we are expecting to obtain from a cost system.

Managers do not necessarily acquire those insights in business schools. Hence, we have the need for new departures in company training, determining which core activities to emphasize which is not always easy or obvious.

Often a company's true strengths are obscured by management's tendency to think in divisional terms and by each functional group's need to see itself as the main source of the enterprise's success. But careful analysis helps identify the few critical activities that drive the company's strategy, its profit figures and its cash flow. This is the approach that must dominate.

5. RETHINKING OUR COST CONTROL POLICIES

One of the first and foremost opportunities an integrative solution will offer is the ability to evaluate companywide our cost policies. Invariably, this leads to the discovery of significant gaps in the cost structure as well as missing or misinterpreted items, thus leading to the rethinking of our cost control requirements.

Starting with the fundamentals, as the economy undergoes huge changes costs become a focal point of attention. But as already emphasized, costs cannot be controlled piecemeal. They need a global plan and this requires an integrative approach to management with:

- Cross-company profitability evaluations,
- Growth perspectives as well as their opposite: downsizing, shedding divisions, and
- An evaluation of cash flow and negative cash flow

In 1990, Chrysler is expected to earn nearly $200 million. But after subtracting the big capital spending bill, there will be an $800 million negative cash flow—a concept, which until quite recently, was not popular but, as companies (and the national economy) have become overburdened with debt, it has become of critical importance.

This polyvalent approach to business management is fairly new. Finance and technology have in common a rush towards new concepts, which however is not a culture in all organizations. For many companies,

- Intelligence about the galloping developments in finance and technology is inaccurate,
- Supplies of needed skills are uncertain,
- Training does not fill the growing gaps in know-how and,
- Most importantly, *adaptability is nonexistent.*

Had such intelligence, for instance, been available, it would have suggested that robotics and AI will eventually narrow the labor-cost advantages signficantly of such would-be auto challengers in Korea and China. GM explored the boundaries of this idea with its Saturn project.* But at the same time, as far as the number of jobs is concerned, sophisticated assembly systems ultimately pose a greater threat to blue-collar jobs than do shipments of imported cars.

Cash flow analyses, profitability studies and cost evaluations are not done in the abstract. They call for a realistic approach to competitive strength, such as the one in Figure 15-3 which reflects on 1989 data and contrasts:

1. Sales per employee (at $1,000), and
2. Pretax profit margins.

Superficially, these statistics have nothing to do with CIM and the AI-oriented concepts that we are investigating—and yet, in reality, they have everything to do with both subjects. A similar statement can be made about marketing results in spite of the fact that so many people fall victim to the easy way out; they equate selling with selling themselves as media personalities.

In some industries, such as computers, a basic criterion is market share in a global sense because the R+D and manufacturing engineering expenditures are so big that a market

---

* See also the opening chapters of this book.

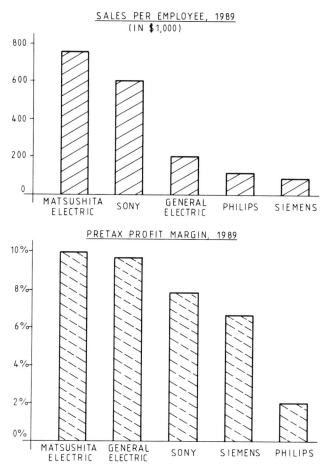

Figure 15-3. Sales and profit statistics for five selected international manufacturing companies.

share of less than 5 percent worldwide is a prescription for financial failure. There is a synergy and a mass effect to be gained from market share.

Reflecting on the market share issue, an article in the *Economist* was to suggest:* "Philips could only scrape together a 1 percent share of the European computer-systems market, less than European rivals like Siemens and Olivetti. The reason? Philip's failure to spot the trend in the market away from minicomputers towards personal computers."

In finance, marketing and technology, the measurement system we develop and its *metrics* should be maintained and dynamically modified. Interrelations between issues should be classified. Costs are a function of the technology we employ, as we steadily weed out manual operations. A full cost system includes:

---

* April 7, 1990.

- The overall cost structure,
- The development or choice of cost methods and standards,
- The cost proper and then the pricing of products and services, and
- The cost and pricing of customer relations.

Computer-integrated manufacturing aims to bring these four issues, which classically belong to separate business sectors of the firm, into the mainstream of technical operations. This requires a cultural change to be implemented, but also networked workstations and databases to make the data available and live at every workbench.

Expert systems are of significant help in interpreting and visualizing such information, but this reference should not mask the need for the proper infrastructure. Quite fundamental, for instance, is the organizational distinction between:

- Profit centers, and
- Cost centers.

As income earners, *profit centers* are directly affected by the pricing structure. Income from *selling* our products and services should be credited to the profit center which has a responsibility to operate productively.

In contrast, *cost centers* survive through budgetary allocation. Cost centers are service providers, oriented toward other company units. Their money comes out of profits before taxes. A profit/cost center structure should above all be streamlined, avoiding the overlapping image reflected in Figure 15-4 that shows the illogical and inefficient organization of many manufacturing enterprises.

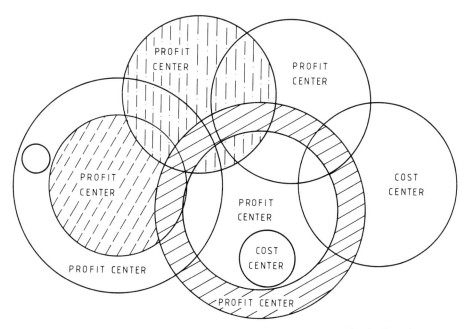

Figure 15-4. The organization of profit centers often results in many and irrational overlaps.

The basic reason for this confusion is management's inability to draw the line and assign profit center responsibility and accountability. Another key reason for the confusion in profit centers and cost centers is the discrete islands approach. By contrast, CIM should permit:

- The gathering of corporatewide financial data,
- A projection on future income and expenditure,
- The making of a budget versus actual evaluations, and
- Transportation expense allocation toward other centers.

As these examples help document, while engineering design and manufacturing perspectives are important elements in computer-integrated operations—and in several ways they are quite challenging—CIM implementation means polyvalent communications as well as streamlined organizational issues. It also means change. It affects the psychology of people and impacts on organization and structure; it requires new skills, attitudes and awareness; and it needs steady performance measurements. CIM is in no way a one-issue channel.

6. THE COST OF STAYING IN BUSINESS

Computer-aided manufacturing (CAM) leads to computer-integrated manufacturing, posing an evident system problem: Who is to integrate the computers that run the factories? An inventory taken at General Motors indicated that there are in the whole company some 65,000 types of program-controlled machines from large computers to small drivers, and they are not necessarily compatible.

This is not the exception, but the rule. From the factory floor to communication networks, large industrial organizations spend a significant amount of their capital investment providing interfaces among incompatible computer equipment, controllers, and protocols. This fact led General Motors to establish uniform standards for vendors who want to do business with the corporation.

While GM's Manufacturing Automation Protocol (MAP) and Boeing's Technical and Office Protocol (TOP) aim to answer the technical side of CIM requirements, less appreciated, but equally important, is the fact that the financial side too must be served through a uniform standard. This is the background of the references made in section 4 about standard costs.

Provided that a sound, accurate standard cost system has been established and is properly kept up, the salient problem becomes one of establishing the proper structure for corporatewide management accounting. This is particularly challenging for national, multinational and transnational corporations, as a valid solution must:

- Accurately reflect the *true cost* of providing our products and services to our customers,
- Permit the pricing of products and services for a fair return on capital and goods profits, and
- Provide a common denominator, a *yardstick* by which productivity achievements, or a lack of them, can be measured.

The financial side of CIM should indicate *cost trends* quickly, making actual reports more useful by showing when and where costs are out of line. It should also assure the basis for a better control over costs, focusing attention on the production and distribution cycle as well as seeing to it that control rests on a *documented, analytical* ground.

Direct expenses incurred by a business segment must be assigned to it. Indirect expenses (overhead) must be minimized. Both operations require a valid *costing* scheme and presuppose a good enough method for analyzing and estimating costs. If the company does not have a formal cost system which has been applied over a number of years and, hence, is streamlined, cost estimates will be rough.

When we evaluate profit and loss—the pricing products and services—we also need to know which costs for a particular service are *fixed* and which are *variable*. This sounds like old accounting, yet it is at the heart of management application to be supported companywide through CIM:

- Fixed costs do not change as the rate of output varies.
- Variable costs vary with changes in the rate of output. Typically, they do so linearly, though this is not always true.

The reason why we encounter a step function in variable costs should be found in the existence of a *semivariable* cost chapter. Semivariable costs do change with the rate of output but they do so by big steps, and then they hold steadily over a range of output fluctuations.

Investment which we have made in plants and in automated machinery (including computers and communications) have to be depreciated immaterial of whether or not they are producing anything near the goal for which they are acquired. In this way they become part and parcel of fixed costs. If they are written off prior to depreciation, the loss has to show in the balance sheet.

It may sound silly to insist on these issues since they have been known to the manufacturing industry since pre-World War II days. Less known, however, is the fact that as manufacturing operations expanded:

- Some of them have been lost from sight,
- Others have been kept on a strictly local basis, cut off from indicating the true cost of doing business, and
- Still others have become too inflexible, which we now aim to correct through the implementation of expert systems.

There is, as well, an issue of coordination. With CIM, we have a unique opportunity to provide the integrative approach most companies today are lacking.

Our cost evaluation will be accurate only when we establish the profile of each source of cost, controlling its *variation*. Companies that do not know their cost for each product or service are taking unwarranted risks and are among the companies least profitable.

CIM solutions should not only reflect, but should also give, a clear picture of the total cost of staying in business. Quite often, management fails to make the vital distinction between:

1. The *cost of doing* business, which practically includes all operating costs, and
2. The cost of *staying in business* which involves the money invested in research, development and new technology.

When the operating costs get out of control—as so often happens—they leave precious little to be invested into the mission of staying in business. One sure way for companies to go bankrupt is to fail in their investment in their own future.

The able handling of the cost of staying in business brings into perspective the product and process life cycle characteristic of manufacturing industries. As the preceding chapters of this book document, this involves a number of critical steps:

- Market analysis and product planning,
- Research and product design,
- Manufacturing engineering,
- Product and process testing,
- Production proper,
- Quality control,
- Sales and distribution,
- Field service, and
- Quality analysis and reliability.

In order to improve upon these successive phases of manufacturing, the foremost companies have turned towards communications, computers and CIM. But dumb computers can offer only limited support. Beyond that point we need to:

- Understand products and processes thoroughly,
- Do competitive research, and
- Introduce intelligence-based methods.

The task is doable, and the preceding 14 chapters demonstrate how it can be done. What I particularly want to stress at this point is that our approaches should not be partial because then they become negative in their contribution to the company's survival.

An integral part of any effort involves whether or not the solution sought after is sound. But it is wrong to look only at the money invested and then try to measure cost effectiveness. Goldman Sachs suggests quite the opposite: Account *only* for new money invested in technology (the individual development cost) as the criterion for living up to evolving standards.

The given reason for this approach, says the investment banker, is that "If you made mistakes in the past, they do not count." But who can tell if we will keep on making mistakes in the present and in the future?

To answer this question we should look at policies, particularly appreciating those companies that do *not* treat computer-integrated manufacturing as a technical problem but look at it, instead, as a way to cut fat and improve business. Such an answer must be businesslike. Considering only the technical issues leads to significant distortions.

This is not the way the majority of companies approach the issue of their own survival. In a recent study, 52 percent of the respondents said that they want technology to help improve management, but only 14 percent of the respondents are willing to use gains in management to justify technology costs. Yet, the *real benefit* involves improving the management process.

At the same time, management improvements cannot be obtained without changes in organization and structure, simplifying and making the communication lines more direct.

That means reducing the number of intermediate layers and improving the computer literacy level of *all* our personnel.

7. CIM AS THE KEY TO MANUFACTURING COMPETITIVENESS

In the 1950s we started with numerical control (NC). But our experience with the numerical control of machine tools did not lead towards a radical rethinking of processes and products, though it did alter some of their aspects. Over the years, the accumulation of knowhow regarding the automation of business processes led to CIM. This happened in the mid- to late 1980s with almost 30 years of elapsed time.

Even during the 1980s we mainly thought of CIM as a better way of using computers in conjunction with flexible automation. The approach adopted included not only computer-controlled machinery but material-handling systems and supervisory solutions, i.e., controllers, as well.

Though the scope broadened and we did move towards integrating islands of flexible automation—including databases and networks—the goal still fell short of the needed approach. As we saw in the preceding sections, the latter should definitely include marketing and finance. This will be the characteristic of the 1990s.

According to the concept the preceding paragraph outlined, we should appreciate that a whole range of goals should be handled. They come into perspective within the realm of CIM in the version of the 1990s, and they include:

1. *Better management efficiency* by eliminating most of a planner's and supervisor's paper work, permitting them to concentrate their efforts on production problems.
2. *Tighter control and utilization of labor* by planning setups in advance and by automating and robotizing the production processes.
3. *A better utilization of equipment,* since production schedules can be planned more efficiently thus reducing downtime due to changes in tooling on rush requirements.
4. *More effective financial control,* since requirements can be anticipated, alternatives evaluated, and performance projected as the time capital budgets are established.
5. *Accurate sales-inventory-production* coordination particularly in all cases where mathematical simulators and heuristics are used.
6. *Improved scheduling and expediting control,* since planning and executing the entire operation are done in an integrative manner.
7. *A better management of raw materials,* by closely following up scheduled quantities, thus preventing overruns and underruns, and by more rapid detection of unexpected manufacturing difficulties.
8. *Improved inventory control* by permitting more accurate and faster advance scheduling, including just-in-time solutions.

9. *Significant quality control improvements* by providing a quality database, enriched with inspection and field feedback information and expert systems to help in quality assurance.
10. *An effective response to market drives* through a close coordination with marketing which should be a mark of distinction with CIM.

One of the major attainments of computer-aided manufacturing should be quick and dependable communication among supervisory, productive and servicing employees in a cross-departmental sense. The network must serve as the focal point for communications: collecting, recording, and distributing all information. This is the essence of integrating discrete applications islands, and of creating a global database.

Sure enough, such claims made in the past failed because the implementation of the plans made did not alter the way decisions took place. To improve upon such performance, we have to have more coordinated decisions and this is precisely where expert systems can make the difference.

Many people within a corporation need access to text and data concerning orders and shipments. These users include distribution managers, sales managers, accounting personnel, production planners and corporate managers. One of the benefits CIM offers these users is immediate access to information on which to base decisions. This includes detailed current and historical data on all orders, products, customers, prices, and inventories.

Such a capability can be further enhanced through AI constructs and simulators, supporting ad hoc queries on exceptions, summaries and correlations developed one time only and interactively displayed. The user can access the needed information with the aid of the appropriate software, thus eliminating the need to go through the data processing department.

Expert systems can also help to audit order-entry activities, stocking-point shipments, and all other contents of the database from any point in the country or outside of it. This, in turn, provides for minimal staffing for day-to-day operations because the user communicates directly with the system through his workstation.

While systems architects will provide the background necessary for designing CIM solutions, as Chapter 16 will show, in reality, the whole organization has the responsibility to keep CIM functioning. This is valid all the way from communicating administrative messages through electronic mail, to the development and use of AI constructs and the refining as well as transforming the ongoing implementation to match changing market requirements.

In this case of greater user involvement, CIM meets two major needs of manufacturing companies that have distributed operations. The first is the accurate handling of client wishes and ordered goods on a timely basis. A company's ability to do this is essential if it wants to maintain its position in a competitive marketplace.

The second need is the realistic forecasting of production requirements to maintain low inventory levels and smooth scheduling. The recent shortages and increasing costs of many resources have placed greater importance on just-in-time control. In addition to meeting these major needs, the able use of CIM can also provide other benefits:

- Forecasting is done on a more actual basis.
- Immediately available displays of up-to-date product status make better management feasible.

- Past and current business can be analyzed by product, geographic area, salesman, stocking-point, and customer.
- Production runs can be optimized and schedules become more effective.
- The price and sales history of our products can be easily reviewed.
- Profit margins by product, distributor, client and a host of other criteria can be easily produced.

Customer service is improved because orders are completed in a more timely fashion and shipment verification can be given at order-entry time. Also a customer's queries about an order can be answered immediately by checking the previously entered order online.

Orders can be edited at the point of entry and, if need be, corrected immediately as they are entered. One system handles the entire procedure, thereby reducing the number of transcription activities and error possibilities.

By linking the factory's information system online to the client's system, orders can be changed easily, immediately updating inventories and checking on committed stock levels each time shipments are made (are expedited or arrive). Errors can be corrected immediately while bookkeeping and paperwork are greatly reduced.

16

Projecting a Rational CIM Architecture

1. INTRODUCTION

Chapter 15 presented the reasons why computer-integrated manufacturing means a longer term vision of the future of fabrication systems. CIM also reflects the changing nature of production activities, its perspective including both the process and the product.

Much of the change for which we have to account is structural. For most of the twentieth century, manufacturing meant mass production for broad markets. Now the goals have changed and as I have already underlined the competitive edge in manufacturing has become *flexibility* and *personalization*. This means:

- Adaptability and variability,
- Shorter, better focused product runs,
- Fast response to market demand, and
- Care for the specific wishes of the individual customer.

The old method practiced until recently, was a hierarchical approach with each step passing along to the next one its physical results, but rarely the rationale behind its *choices* and *decisions*. This ended with the enlargment of the time frame, creating misunderstanding, generating delays and leading to high costs.

In contrast, with CIM, we aim not only to facilitate operations within a given step, but also to support cross-step *knowledge transmission*. Our goal is for consistency to be maintained in a companywide sense. This is why the CIM architecture we will adopt must be supportive of a cross-step framework bringing together:

- People,
- Systems, and

- Activities.

At the same time, to integrate everything done in the company we need a comprehensive model of organization and structure, encompassing all assets and linking them to each other.

Far from being a theoretical exercise in system integration, this strategy has specific goals to meet particularly adapted to ongoing market wishes and drives. These often constitute a moving target but in the background they include:

- Sophisticated services,
- Support of greater comfort,
- Styling and novelty, and
- Costs and benefits.

One of the particularities of these demands today advanced by the market is the urgent need faced by industrial organizations to develop the expertise that permits them to deal with future events in product design and its fabrication. This has to be done without losing the focus on sustained *profitability*.

Both internal and external information exchange requirements are therefore imposed, and they have to be fast and reliable. Around this central concept of a system which must be:

- Physically distributed, but
- Logically integrated

the corporate function agents operate. This involves market research, R+D, product planning, design and engineering, manufacturing, just-in-time inventory, cost control, profitability analysis and cash flow.

These perspectives define the architecture of evolving manufacturing facilities and by consequence of CIM under its definition in the 1990s. To show this, in the present chapter we will follow some of the most important architectural requirements for the successful implementation of an efficient computer-integrated manufacturing solution.

2. THE ARCHITECTURE OF THE FLEXIBLE MANUFACTURING ENTERPRISE

The global market demands greater flexibility; an accelerated pace of innovation, and thus faster product introductions; stable scheduling, yet shorter production runs; and a continued refinement of the associated financial functions. Within this landscape, corporate success depend on countless, highly interdependent decisions by hundreds or thousands of individuals, with a number of factors contributing to the pace of development.

In Chapter 15 we saw that organizational perspectives dominate in spite of the fact that public attention today is carried more towards technology. Even with computer systems spread around the globe at corporate headquarters, engineering centers, factories, and clients' and suppliers' offices, geographically dispersed manufacturing companies can be no better than their organizational infrastructures are.

Decision makers and agents interact with each other through a shared, distributed base of knowledge. This has to be tuned to accelerate the flow of products and to coordinate all related activities.

This increasingly complex setup that ties together people and machines can only function successfully if there is an enterprisewide knowledge system that builds on and unifies three technologies:

1. Globally networked data bases and a myriad of workstations installed at every post the company operates
2. A common, corporatewide semantic symbolic model of the enterprise able to facilitate information exchange
3. A means for linking the business partners online through intelligent agents, able to create a cooperative environment.

Some researchers are calling the latter *anthropomorphic agents,* giving as examples sales analyzers, just-in-time inventory controllers, client-order takers, distributed schedulers, market data filters, cost controllers, and quality assurers. The background reference is of course that these are not people, but knowledge engineering constructs.

Such a system could eventually work if the knowledgebank of the manufacturing company is a systematic and noncontradictory aggregation of complete semantic models. Creating an aggregate of the enterprise, such models would map its:

- Physical resources such as product inventories, facilities, equipment and raw materials,
- Human and capital resources, and
- Informational and knowledge engineering capabilities including product designs and manufacturing processes.

They will also address and integrate the organizational structure and the personal responsibilities as well as those accountabilities and operating procedures the organization defines.

Such a scenario may seem quite futuristic and no doubt it will take more than this decade to be completed. But if we are today to adopt *a new architecture* for the manufacturing industry, such an architecture will have to be forward looking; otherwise, it will be overtaken by events.

A modern system architecture has to account for the fact that decisions in a manufacturing environment, which has only slightly improved over the years, are not made easily because of:

1. *Information complexity.*

An example is our effort to automate data capturing. Bar codes permit us to solve the input problem up to a point but they do not change the information handling process. On the contrary, they inundate it with more data than it is designed to accommodate.

The crux of the matter is that simply having information available is insufficient. This information has to be digested in order to create a successful design, manufacturing, sales and distribution system. A company needs to modify and enhance its basic infrastructure to meet evolving requirements, and this presupposes the ability to break with the systems and procedures it employed in the past.

Information complexity can be reduced through the appropriate experimentation. In Figure 16-1, this is shown by means of a virtual factory concept. The experimental line of reasoning provides:

- Fast feedback,
- Uniform, high-level access, and
- A testing ground for automation research.

In the past we lacked the knowledge and the tools to implement such solutions. Today we have both, hence, there is no excuse for not using them. Besides, their absence leads to the fact that in many manufacturing companies:

2. *Information is not being properly utilized.*

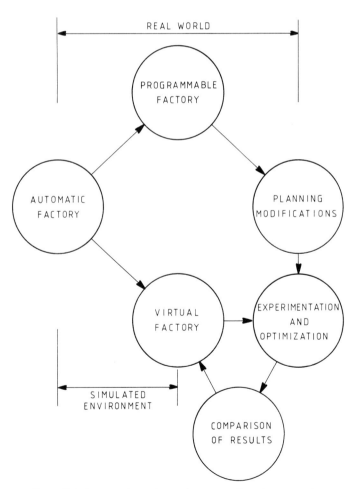

Figure 16-1. A sense of experimentation on manufacturing automation.

Just because data is overproduced does not mean that it is also properly employed. The right exploitation of databases starts with implementation planning, which calls for functional requirements analysis and specification as well as customization and testing of programs and, without doubt, system implementation and training.

For each service to be rendered, it is necessary to prepare a technical description of the work to be performed, the personnel to be interfaced, and the information products expected to be delivered. Also to be considered are the estimated time frame, anticipated costs and expected results.

In order to improve upon the current state of affairs, we develop and use expert systems. Knowledge engineering solutions also help in overcoming the next handicap:

> 3. *The present lack of dynamic enterprise concepts.*

As we approach the year 2000 we find that we have too little expertise available to do the jobs that need to be done. Scarce expertise is a severe problem to many industries. The skills of the very few experts we have are needed in 100 different places.

Lack of enough skill has consequences are not always appreciated. The most important is the fact that people who lack the required expertise—including CEO—act in a retrograde fashion or take unwarranted risks. This is the bravado attitude which rarely if ever pays dividends.

Since we cannot divide up physical people, the solution is to multiply their skills and experience through AI. Clear-eyed corporations are doing just that and the foremost among them build knowledge engineering into their system architecture.*

However, it is advisable to remember that a new manufacturing architecture cannot be made in the abstract and neither can it be designed detached from present-day systems and procedures. This is valid no matter how irrational the latter may be.

Beyond any doubt, we need to estimate the total impact the installation of the new architecture will have on the organization. We need experts to assist in this effort by defining software and hardware requirements (for development, testing and operation), interfaces with existing systems, necessary conversion tasks and personnel training needs. We also need schedules for phased implementation from design to system testing to cutovers.

3. CONCURRENT ENGINEERING, AN EXAMPLE OF CIM ARCHITECTURE

A manufacturing company that intends to assure itself a position of permanence as a partner and a provider of *solutions,* not just products, must see to it that the decisions taken by its top management possess six key elements:

- A specific market-oriented strategy that includes definitions, not just descriptions,

---

* See also D.N. Chorafas, *System Architecture and System Design,* McGraw Hill, New York, 1989.

- An architecture based on this strategy which can be used as the frame of reference,
- A properly tuned development program, including all aspects necessary for software integration,
- Emphasis on knowledge engineering approaches, and the appropriate skills to implement them,
- Supporting services for the realization of computer integrated manufacturing, and
- Centers of competence to test the solutions which will be spreading over its global operations.

No two CIM strategies are the same. Each has aspects that are different from that of other companies. Said the executive of a major computer vendor: "When we started looking into the CIM problem in detail, we identified three areas: Projecting the product and the industrial process; the management of resources; and industrial automation/robotics." From one company to another, there exist variations in each one of these areas.

A case in point is the XYZ Corporation. A thorough study conveyed to the company's management the message that there was little integration between projecting:

- Products and processes, on the one hand, and
- Resources planning, on the other.

As a result, this company faced and often adopted diverse solutions in the automation of single operations. Finally, management became aware that the traditional approaches were fundamentally different among themselves and therefore heterogenous.

In the area of product and process design, what was particularly noted was the need to integrate, in a flexible way, the results of laboratories established on three continents. All of them belonged to the XYZ company but they operated as independent business units.

This awareness made it mandatory, as a first step, to understand what it takes to integrate discrete islands, tying together different types of *thinking* in the engineering world. The same reference was found to be valid for resource handling, including production planning and operational control in the factory. Then came the need to integrate the latter two worlds vertically.

Another firm that specialized in electromechanical devices spoke of the need to develop *concurrent engineering* solutions on a cross-country basis supported through CAD and expert systems, and including:

- A preliminary design,
- A detailed design,
- Engineering analysis, and
- Manufacturing engineering.

The concept of concurrent engineering implies that instead of developing products step by step, the foremost manufacturers want designers, engineers, parts and component vendors, their own buyers and their own salesmen to work *in tandem*. This can be nicely done through networked workstations—provided the appropriate policies, concepts and software are in place.

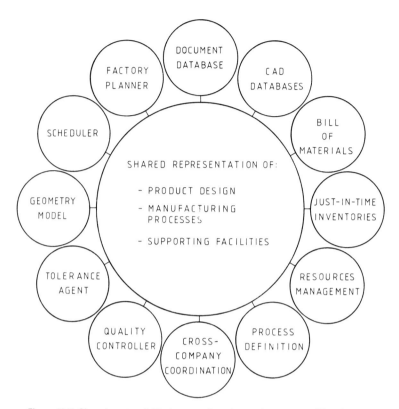

Figure 16-2. Shared responsibility for many functions to be automated in a factory.

This type of corporatewide coordination was found to be at the basis of the necessary infrastructure in coordinating engineering and manufacturing across different countries and different cultures. An integrative approach to this solution is shown in Figure 16-2.

It will be appreciated, that the core is composed of a shared representation of product designs, manufacturing processes and supporting facilities. The twelve agents around this kernel are essentially satellites to the core which have to be highly personalized—and this can be effectively done through expert systems.*

Hence, we have the need to establish a *metaknowledge engineering* discipline that permits us to provide inheritance in the different functions mapped into the personalized model. Such an approach should work in conjunction with each factory's (and the end user's) specific requirements and characteristics.

This model of CIM architecture is inherently distributive with the different nodes interlinked through *liaison agents* that assure that everybody affected by a certain action (a plan, a design, a decision) gets notified. Such an aggregate helps provide a natural *metaphor* for mapping the whole enterprise into the network and its data bases—provided that the proper prerequisites are observed.

---

* See also D.N. Chorafas, *Knowledge Engineering*, Van Nostrand Reinhold, New York, 1990.

One of these prerequisites is that automated jobs are handled through communications and computers, not 50-50 through technology and paper shuffling. Mixing paper and CIM will end by:

- Paying computer costs which are high
- While getting paperwork efficiency which is low.

This message has not been properly understood in all quarters. Many organizations have installed CAD/CAM systems in their design engineering and manufacturing departments without reaching the improvement of productivity which was foreseen. Because the paper-based skills still dominate, their designers spend much of their time producing a greater quantity of projects on paper than on the computers.

This of course runs contrary to the strategy of integration which consists of recognizing the diverse nature of different operations, subsequently bringing them together in a functional manner. When this happens, we can use a logically homogeneous model, and proceed with integration in an application-independence sense.

The idea is to provide a shared knowledgebank as shown in Figure 16-3, keeping in perspective that the heterogeneity of a given system does not only reveal itself in internal applications. When the salesman visits his clients, he invariably finds diverse solutions as these clients buy hardware and software from many manufacturers and the latter use incompatible systems. And the same is true of the purchasing agents.

4. TUNING THE CIM ARCHITECTURE TO THE PRODUCT LINE

In a study focusing on new products, Booz-Allen & Hamilton looked at 700 US manufacturing companies operating in America and in Europe and their 13,000 "new product" offerings. This study documented that over a five-year period, only 10 percent of those products were truly *original,* one third of them being in electronics.

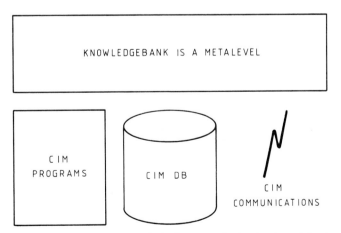

Figure 16-3. CIM solutions require both an operating level and a metalevel.

More than half the studied companies failed to introduce any original products at all, and it is not surprising that the same firms also failed to keep pace with manufacturing, marketing, and management techniques. This is quite often the case: Product failures and process failures go hand in hand.

That emphasis should be placed not only on new products but also on new processes is documented by the fact that automated production has helped Japan dominate markets ranging from autos to consumer electronics. To make a television set, a Japanese firm typically uses 30 percent fewer components than a European or American firm does, and it does so by automating up to 80 percent of production—far more than its western rivals do. It requires:

- 1.9 man-hours to produce a television set in Japan,
- 3.0 man-hours in America,
- 3.9 man-hours in West Germany, and
- 6.1 man-hours in Britain.

No wonder that American and European consumer electronics companies are now searching for ways to revive their past dominances. Solutions, however, are hard to come by because management does not take the proverbial long hard look. What these companies invariably find is that much of what fails them lies within their own management practices.

A good example is given by one of the ironies to be found in manufacturing automation. Computers help product designers and engineers, but it somehow seems that:

- Much of what engineers do is busywork
- They are constantly revising a given product and its parts because the various design teams don't know what the others are up to.

More than 50 percent of engineering design changes being made are avoidable if better coordination is established. Hence, we recall the concept of concurrent engineering of which we spoke in Section 3.

The pivotal point in a concurrent engineering environment is a knowledge engineering solution with both integrative and switching capabilities as shown in Figure 16-4. Around such a kernel is a constellation of data bases and users:

- Many of them supported through expert systems, and
- Practically all of them having weeded out paperwork with its costs, errors and delays.

The basic premise is one of integrating engineering design premises focusing on product perspectives with the development of efficient manufacturing processes down to the level of facility layout. This means integrating machines, work areas, process plans, product characteristics and market demand.

The benefits that could be derived from concurrent engineering are electrifying the American manufacturing industry. Good implementation examples are:

- Sikorsky Aircraft's Igor, and
- Schlumberger's Electronic Bill of Materials (EBOM).

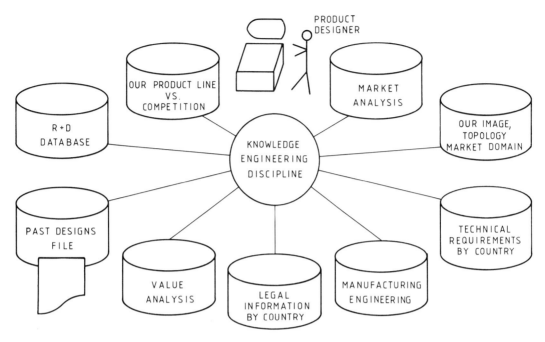

Figure 16-4. Concurrent engineering for product design with expert support systems.

Sometimes called engingeering data management or electronic design management (EDM), the concept is that of slashing the *time to market*.

Slashing the time to market as well as increasing engineering productivity have been goals of computer-aided design since the start, but paradoxically, in many cases, CAD increased the flow of paper. That is due to two reasons:

- Different brands of CAD systems cannot easily communicate with each other, and
- Companies have been slow or unwilling to change their practices.

As a result of the heterogeneity in formats, when sending a part design to other departments or to another company (the client, the supplier, the controlled subsidiary), designs must be printed on paper, then manually fed into each recipient computer:

- The process is slow, costs money, and runs the risk of human error.
- All too often, a revised design arrives after an assembly containing the part is in production.
- An engineering change order must be issued, and the fix is not only costly but it also upsets production plans, requires retooling, and creates delays in getting to market.

The problem is so prevalent that Ford invested seven years and $77 million to develop a proprietary solution, called Worldwide Engineering Release System (WERS) which became operational in 1989. "It was an approach we felt we had to implement in order to do our product planning on a global basis," said Charles L. Bambenek, a WERS project executive.*

Other companies have taken a similar approach and specialists suggest that CAD/CAM integration shows return on investment ranging from 190 to 355 percent. This has been achieved through architectural approaches along the lines of rationalization which we have been discussing.

The lesson to learn from these references is that technology has the potential of changing the way we look at things. The trend toward networking flexible solutions has gained momentum in manufacturing. Automation has not only brought a reversal in the traditional cost ratios such as those of direct labor and direct materials—of which we have spoken on several occasions—but also in our way of designing systems.

As it cannot be repeated too often, the market demands personalization and innovation. This has a great impact both on our products and our processes. The speed with which we implement change is crucial to profitability. But to apply a faster turnaround time we must:

1. Change the culture of the engineering and manufacturing people
2. Provide better coordination between R+D and production
3. Reorganize all manufacturing operations according to the new architecture we should design.

Turnaround time can be shorter if we keep with smaller units that work together in *a federated way;* we should avoid mammoth factories. Manufacturing companies are steadily switching from large centralized parts manufacturing and assembly plants to smaller, dispersed production entities, and they watch the individual profitability of each unit.

Because of the impact of technology, NCR, for instance, says that within 10 years the company increased its sales per employee by 280 percent. Yet, that still did not come close to matching IBM, which boasts sales per employee about twice as high as those at NCR.

Sales quotas have been basically applied to marketing while the factory worker was subjected to time and motion studies. However, the dual effect of independent business units and robotics changes this frame of reference. The personalization of products makes feasible production quotas similar to those used in sales.

At the production floor, knowledge engineering and information technology also provide the necessary infrastructure for a better follow-up of activities taking place, assuring:

- Shorter lines in making goods,
- Nonlabor intense operations,
- A sharp reduction of buffer stocks between manufacturing stations, and
- A resulting cost control capability.

In other words, not only do robotics displace labor and alter the classical manufacturing chores, but significant cost improvements can also be realized by computer-based production planning and control based on knowledge engineering. This reference includes individualization of performance, CAD and CAM coordination, optimal scheduling, JIT, quality control, and purchasing activities.

---

* *Business Week,* September 3, 1990.

5. OBTAINING BETTER COORDINATION— FROM PRODUCT DESIGN TO PRODUCTION SCHEDULING

The emphasis on knowledge engineering and the results obtained so far have made us realize that in the past we did not know how to schedule our resources very well. Today there are many more alternatives to choose from, but at the same time:

- Flexibility requires decision timeliness, including experimentation time
- Solutions to the problem of complexity with production decisions done on a much faster time scale.

This requires first-class coordination that must take place not only in a given factory, or even in factories around the same city, but also around the continent and around the world.

A greater degree of coordination can be obtained only by overcoming the production hurdles that exist in many companies. For instance, Hewlett-Packard has created a *digital standard* for describing the 150,000 designs it has accumulated worldwide:

- Permitting the creation of an engineering database, and
- Making it possible that all CAD/CAM workstations use the same text and data.

Manufacturing companies have finally become keen on keeping databased engineering documents in good condition, updating this *multimedia* information often so that no outdated references foul up a design. And the compound electronic documents created in the design department can be used to run engineering analyses or to drive factory computers.

A sensible information architecture for a manufacturing company will see to it that the solution we adopt:

- Lets every single designer see everything around him, and
- Enables everybody in the downstream manufacturing process to take a look at what is being done upstream.

This saves a lot of errors and makes a much better coordinated environment. Such approaches are crucial to concurrent engineering which, as we have seen, involves teams from various formerly watertight departments—including after-sales service and outside suppliers—who cooperate to develop new products and processes.

The foremost manufacturing companies are perfecting programs called *frameworks*, which make existing software work together and communicate with databases. Frameworks impose a common data structure on programs that deal with every element of design.*

---

* At least this is the intention. Current results from concurrent engineering applications suggests that things are not that perfect. Hence, it is wise to first test frameworks offered within one's own environment.

Not only must access to distributed databases be facilitated, but also the type of queries accepted must be ad hoc and permit the handling of vagueness as well as uncertainty. We have spoken of this in Part One and we have also provided specific application examples, including Nissan Motors and Hitachi.

Whether in America, Europe or Japan, the aim of these applications is the realization of an efficient support system for planning and management. Advanced architectures in the manufacturing industry have two significant features:

1. The capability for multimedia information such as figures, data, text, and various attributes as well as images related to them.

This is realized by utilizing relational databases in conjunction with a knowledgebank where the schema information is stored.

2. An intelligent interfacing capability with a natural language, where complex sentences can be interpreted.

This also provides a function for handling polyvalent relations, and infers solutions that are not stored directly in the database.

The solutions we provide must account for factors such as information complexity, scarce expertise, decision complexity, the need for timeliness and coordination. Failure to do this will result in barriers to the effective implementation of computer-integrated manufacturing.

Any effective CIM policy will broaden the application of computer integration to the product's manufacturing life cycle including design, planning, production, sales, distribution and field service. The challenge is not only to integrate:

- Products,
- Processes, and
- Resources,

but also how to upkeep the system in the face of changes that take place in sales, engineering, and manufacturing, as well as in the administration. The goal is that of integrating the whole enterprise—not just some of its parts.

To solve the problems of enterprisewide coordination, companies create computational theories that enable them to perceive, think and act intelligently—but in a way that can be implementated on a computer. To learn how to study and develop solutions, the domain in which we work does not have to be very complex, but it still can present challenges not easily answered through classical approaches.

In essence, we are talking about a class of problems that are *not* easily solved, leading us to provide a layered approach to business assumptions:

- At the corporate level, authorized users can set values for these assumptions.
- At the independent business unit level, an authorized user from each business unit can override corporate settings for any assumptions—provided he justifies his act.

The information utility must see to it that all projects by all users in the business unit pick up these new values. If new values are not entered by a business unit, corporate values will be used with the business unit inheriting the characteristics of its parent.

At the project level, assumptions can be changed by a user for a single scenario. For a corporate database, these settings must be saved along with the project, creating a *corporate memory facility*.

Throughout this process we must understand the implications of our acts, rationalize our approaches, and simplify our ways of doing things. This is what the Japanese have done very well; let's learn the lesson.

There is no use in applying technology in a complex way. Individual situations do not really need to be so complex and if they are they can be simplified through AI. It has taken a very long time for people to understand what the implementation of knowledge engineering is all about.

Yet, even for the foremost companies, the knowledge revolution really started in the late 1980s; and there are many companies for which the knowledge revolution has not yet arrived. Among these will be found the firms in greatest difficulty in the 1990s. This is what I predict.

6. OBSERVING SECURITY REQUIREMENTS

A rational CIM architecture should both define and implement security clauses. This requires both encryption algorithms and the proper definition of authorization and authentication. Two types of authority affect access control:

- Authority relating to people, and
- Authority over resources.

A security administrator issues user access rights to perform operations. Such an assignment is part of the organizational domain that defines users who can engage in access services. This may be unlimited (networkwide) or it may be limited to a particular section: a factory, a design office and so on.

Within the realm of security architecture, the security administrator can only give access rights to a person occupying a position within his or her organizational domain. In contrast, the resource domain defines the set of operations of resources to which the security administrator has authority to grant access rights.

A source of authority over the use of resources or assets is vested in their owner. Some file systems, for example, consider the creator of the file the owner, who can issue access rights to other users: adding, updating, as well as deleting the file. But an industrial system does not operate that way, making it necessary to distinguish between full control and limited control over access rights.

Within this perspective, applications and data should be treated in the same way as other company *assets* are treated, with access rights defined as formally as they are defined for monetary values. The authority for use of resources is typically established through a chain of formal decisions:

- The possession of an access right does not automatically permit the holder to transfer it to another user.

- The authorized acts of both the access user and the security administrator can be traced back to the formal decisions of the delegation of authority.

Commercially available access control systems recognize this fact and present the necessary utilities able to support the lines of authorization and authentication defined within a specific context.

Emphasis on these issues is justified by the fact that security has been one of the weaknesses of many automation systems, the more so when the need for it has not been considered since the drafting stage. A bootstrapping-user ID is one of the risks, and though it can be effectively monitored not all manufacturing companies see to it that this is done.

Given that all technical information is databased, what prevents a larcenous clerk from accessing confidential product design data? In most companies, the answer is nothing. Yet, through the proper controls such a risk can be:

- Limited in magnitude,
- Reduced in likelihood,
- Made more difficult and more detectable.

Each one of these aspects plays a role in overall security as pass-through capabilities are enlarged by means of system integration. Let's always recall that the most important part of our CIM program is the development of such an integrative capability, covering all organizational needs—from the office to the factory.

The security risk can be limited in magnitude by placing upper limits on data base access authority. However, as far CAD/CAM and CIM are concerned, such a risk can only be limited, not removed; it can be confined to well-defined situations monitored by personal supervision.

Lessons can be learned as well from the banking industry with its practice of daily reconciliations and regular audits. These are intended to assure that if a clerk does make off with a handful of notes the loss will be rapidly detected and traced to its perpetrator.

In a similar manner, the actions of a security administrator can be logged and should be regularly monitored. Even if the administrator can turn off the logging, it is possible to assure that this act too is recorded securely, for instance on write-once read-many times (WORM) optical disks.

Sound policies and internal control principles require that resource ownership be unique, deriving from delegated authority down the management structure. Access control procedures must take into account how much freedom users should have in relation to the information resources with which they work.

A similar approach should be undertaken for interfacing control modules as well as databases, design engines and robots. This is particularly true about modules offering support functions in the decision phase, as well as in product and process development.

Policy decisions are needed to constrain the users of database resources to act in ways consonant with the interests of the company. Design engineers and industrial engineers must recognize the force of these decisions as they have been made in accordance with the authority structure of the company:

- A policy decision made by the marketing director is not binding on the manufacturing division unless the board has confirmed it.

- Similarly, access control policies have to be made by the appropriate authority who restricts the giving of access rights.
- This is all the more important as access control policies affect the way day-to-day operations are carried out. Hence, they should be made about positions, not just people.

Nevertheless, given the impact database access has on personal productivity, the need for formal modeling becomes evident. A security model will make it feasible to experiment on rules and their consequences. It will also permit us to formalize better the way in which security regulations are described.

An expert system enacted-security model helps provide a precise specification from which implementation can be carried out. It can be used to validate policies by assuring that consequences will actually be studied prior to formalizing specific rules:

- We can use simulation to explore the consequences of possible access control policies.
- We can also employ an AI model to find security weaknesses in existing access control systems.

This will permit streamline regulations for the control system as a whole and will determine whether undesired consequences could follow. The level of assurance reached should not put a low ceiling on productivity.

The application of a security model will also permit a focus on the construction of a logically homogeneous policy, where rules can be applied without twists and exceptions. This is important as both computer and robot manufacturers often leave each individual factory to reach such decisions with little coordination among themselves.

7. CIM ARCHITECTURE AND MARKET COMPETITIVENESS

Computer-integrated manufacturing calls for significant investments and it can only be justified if we obtain valid results. The No. 1 goal we should be after is market competitiveness which calls for an enterprisewide implementation supporting customer consultancy and enhancing communication within our firm. Distributed AI solutions:

- Promote communication by helping to integrate and coordinate management processes on different organizational levels and in different places, and
- They provide for filtering and transforming information flows, doing so automatically in an intelligent, efficient and reliable way.

The concepts and applications we have seen in this book create a *knowledge network* that constitutes the vision of the business of the future. A number of practical examples have demonstrated the intermediate steps to take to get there, with three main points of reference:

1. An order-processing cycle.

286 CAD, SCHEDULING, MANAGEMENT AND CIM

This ranges from sales proper to order handling by computer, industry production planning, quality control and products ready for use.

2. A shortening of the time needed to develop and market new products.

As we have seen, concurrent engineering accelerates and speeds up the development cycle—a very important part of market competitiveness.

3. The computer-integrated enterprise as a whole.

This is a goal tough to reach but doable. To respond to this challenge, the foremost companies have instituted an across-the-board approach with knowledge engineering.

According to the foremost manufacturing companies, AI is becoming embedded in traditional systems development including software, hardware, databases, and networks. Knowledge engineering approaches will dominate all the way from product design to manufacturing and marketing. In essence, AI and DP will merge as only hybrid systems can respond in an effective manner to all of the ongoing requirements.

As stated in the foreword, the fully intelligent enterprise is an unattainable idealization, since it seems to be impossible to give an exhaustive set of attributes characterizing every single operation in its finest detail. But organizations are artifacts that can benefit from modeling and expert system support, as these help in spreading logic and experience through the corporation.

In the sense of a methodological and consistent approach, able solutions to knowledge representation and inference techniques are realizable and practically useful. By and large,

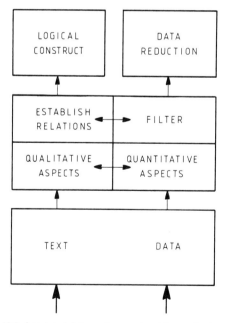

Figure 16-5. A text and data environment and its required supports.

AI models are not aimed to replace but to extend and to complete the existing experience in a manufacturing enterprise.

As we have seen in the 16 chapters of this book, there are many subjects of importance in AI implementation—all with innovation potential—and they can be found in R+D manufacturing and sales. Integrative marketing, product innovation and the production processes are regarded as some of the most promising strategic applications within the modern organization, as they lead to a high performance model.

Both qualitative and quantitative variables play a central role within any company and its environment. Therefore, intelligent reasoning mechanisms can offer worldviews for measurable (quantitative) constraints as well for qualitative expressions and estimates. Figure 16-5 presents an example along this line of reasoning:

- The input may be in the text, the data, or both.
- The aim involves logical constructs as well as data schemes.
- The handling of the text leads to the establishment of relations; the data is subject to data filtering.

The filter may include statistical tables, an algorithm, Bayesian probabilities, or possibilistic expressions. It may also be a set of constraints.

Constraints help establish a consistent and complete state description of the company and its environment from incomplete, inconsistent and fuzzy information. Such a transformation requires facts and relationships as well as a base for the diagnosis of operations and weak-point analysis. Just as helpful is the simulation of different strategies for integrated operations as discussed in the present and in the preceding chapters.

Solutions have to be provided without building mammoth software systems the way they have been made in the past. Therefore, in the preceding sections, I have underlined the importance of a modular approach held together through a properly established CIM architecture.

Solutions have also to be implemented without huge investments in support software as required by traditional approaches. Table 16-1 shows some examples that dramatize the growth in size of classical software we experienced from the 1950s to the 1980s. It is simply not possible to continue along this path.

Solutions have to be constructed in a way so as to deliver the needed services while containing the support size. Answers have to be provided in a qualitative and not just

TABLE 16-1* Growth in Size of Classical Software From the 1950s to the 1980s

System	Operational Code	Support Code
Sage (1950s)	100,000	500,000
Air traffic control (1960s)	200,000	1,000,000
Safeguard (1970s)	750,000	2,000,000
Norad and Spacecom** (1980s)	2,000,000	25,000,000

* *Computer Magazine*, January 1989.
** Includes space defense operations, intelligence data handling, air defense, ballistic missile warning, communications network control, and command and control.

quantitative manner, accounting for synergy and coordination and also delays, as well as interlocks.

The effective integration of key features necessitates the automatic interpretation of information elements in internal and external databases; it also involves the processing of issues containing vagueness and uncertainty—such as the forecasting of market volume by region and type of product:

- An early warning system is another example of needed coordination but also of possible interlocks.
- Weak-point analysis and the whole process of quantitative reasoning including knowledge-based diagnosis are other similar examples.

The proper CIM architecture would handle the problems to be solved by calling on the appropriate software within the domain knowledge and predefined inference schemes. Through knowledge engineering solutions, software will be available for assessing the results, including consistency checks, a comparison of *a priori/a posteriori* data, evaluation of logical schemes and so on.

These references look like details yet they are the cornerstones of improving communications with the customer base and within the manufacturing enterprise. This has to be done all the way from market analysis and sales results to product innovation, production planning, the control of quality as well as the after-sales service. Though necessary, consistency at the top management level is not enough; God is in the detail.

Index

Index

Accident Management Expert System (AMES), 81
Action groups, 18
Advanced Ground Vehicle Technology, 126
Advanced Reasoning Tool (ART), 62, 63
Advanced Telecom Research (ATR), 136
Advanced Test Accelerator (ATA), 83
AI constructs, 93, 225
    diagnostic, 226
    networked, 49
AI models, 287
AI tools, use of, 147
Alarm Filtering System (AFS), 84
Alarm Processing System (APS), 84
Algorithms, 76
Analyses, mathematical, 41
Apple Computers, 15, 260
Architecture, knowledge-based, 61
Artificial intelligence (AI), 27, 44, 67, 90, 165, 237
    implementation of, 46
    in manufacturing, 46
    leadership in, 45
Asset management, 101, 102
Atari, 233, 235, 237, 238
AT&T, 12, 48
Automated Cable Expertise (ACE), 48, 49
Automated system, 5
Automatic Camerca Recognition System, 35
Automatic factory, 7
Automatic guided vehicle systems (AGVS), 129
Automation, computer-based, 22
Autonomous vehicle technology, 127, 130

Backbone design, challenge of, 75
Backbone network, 75
Bambenek, Charles L. 280
Barnett, Steve, 234
Baroid, 115
Battelle, 154
Bechtel Corp. 57, 58, 67, 84, 85, 132
BBN expert system for network design, 73
Bell Laboratories, 128, 215
Bill of Materials (BOM), 92, 98, 141, 181, 200
Boeing, 264
Boeing AI center, 67
Bolt Beranek & Newman, 59, 69, 72, 74, 77–79
Booz-Allen & Hamilton, 277
British Airways, 36
British Coal, 68, 69
Buick City, 216–18
Burroughs, 236

CAD, 30, 141, 142, 144–50, 155, 156, 165
    intelligent, 160, 169
    second generation, 142
CAD expert systems, 158
CAD workstations, 29
CAD/CAM integration, 280
CAB/CAM systems, 277
CALIDA, 246
Candler, Asa Griggs, 240
Capital assets, 101
Capital Assets Expert System (CASES), 100–04
Carnegie Mellon University, 94–96

CASES user guide, 103
Causality, 198
Chrysler Motors, 162, 258, 261
CIM architecture, 276, 283, 288
Citibank, 240
Classification, 117
Cognitive engineering, 149
Cognitive science, 113
Colgate-Palmolive Co. 37
Command-and-control mentality, 33
Commodity structure, mixed, 57
Communications, 38
Communications solutions, 30
Compaq, 232, 233
Complexity, 73
Computer-aided engineering (CAE), 57
Computer-aided manufacturing (CAM), 30, 143, 264
Computer-integrated enterprise, 179
Computer-integrated manufacturing (CIM), 23, 27, 183, 252–54, 256, 261, 263, 264–68, 271, 275, 277, 279, 282, 285
Computer programs, 17
Concurrency, 38, 167
Concurrent engineering, 275
Configuration, 92, 94
Configuration process, 98
Configurer, 73
Connection Machine-2, 7
Connectivity analysis, 173
Connectivity analysis, algorithms, 73
Consistency checking, 92
Constraint satisfaction, 203
Constructibility checker, 57
Consultation functions, pre- and post-, 103
Continental Airlines, 87
Controllers, artificial intelligence enriched, 6
Control optimization, 169
Corporate memory facility, 283
Cost centers, 263, 264
Credibility, 78
Credit Clearing House (CCH), 238, 239
Customer satisfaction, 78

Daifuku, 146
Dai Nippon Printing, 146
Danzig, Georg Dr. 186
Database facilities, distributed, able management of, 80
Database management systems (DBMS), 71
Databases, federated, 80
Database structure and meaning frames, 66
Data filtering, 212
Data General, 22, 23
Data transfers, 111

DEC, 12, 37, 74, 87, 88, 90–95, 97, 167, 190, 243, 258
DeCastro, Edison D. 22
Decision Expert, 88
Defense Advanced Projects Agency (DARPA), 129, 130
Department of Commerce, US, 4
Department of Energy, 86
Design, dynamic, 161
Design, logic-level, 51
Design techniques, symbolic type, 161
Development process, 103
Development team, 32
Diablo Canyon Nuclear Power Station, 84
Diagnostic process, 225
Dial tool kit, 115
Direct labor, cost of, 3
Direct Labor Management System (DLMS), 257
Dispatching, 180
Distributed deductive databases (DDDB), 119
Distributed negotiation, 211
Domain experts, 103, 104
Drucker, Peter Dr. 13, 18
Dun & Bradstreet Business Credit Services, 238

Electric Power Research Institute, 85
Electronic design management, 279
Electronic Document Interchange (EDI), 37
Electronics products plant, cost in, 3
EMES (Energy Management Expert System), 53
Enterprise Map, 88
Entropy, 55, 56
Episodic memory, 120
Experimental Test Accelerator (ETA), 83
Expert Configurer (XCON), 74, 90, 93, 94, 96–98, 100
Expert network performance analysis module, 74
Expert Selling Tool (XSEL), 74, 90, 92, 93, 97
Expert Site Preparation Tool (XSITE), 90
Expert systems
 analytical, 50
 application fields of, 50
 configurator, 92
 data-driven, 63
 diagnostic, 70, 230
 fault isolation, 52
 implementing, 13
 in manufacturing, 28, 45
 in marketing, 47
 in production planning, 179
 interactive, 102
 justification for implementing, 55
 second generation, 133
 synthesis-oriented, 94

Expert systems
  ACE, 227
  APEX, 227
  ARBY/NDS, 227
  ATES, 82, 83
  AUTOMECH, 227
  CAF, 80
  Care, 89
  Class, 89
  CLIPS, 62
  COMPAS, 79, 80
  Control Value Selection Expert System, 58
  CRACKS, 68
  DART, 227
  DELTA/CATS, 227
  DESIGNNET, 72, 74
  DESIGNWORLD, 224
  DIMOS, 226
  Dipmeter Advisor, 153, 154
  EDS, 224
  Email Browsing, 88
  EXMAR, 68
  Fan Vibration Advisor, 58, 86
  FIES (Fault Isolation Expert System), 53
  FORBIN, 224
  Gatekeeper, 87
  GENTLE, 64, 66
  IDA (Intelligent Database Assistant), 80, 246
  IDT (Intelligent Diagnostic Tool), 90, 227
  ILOG (Intelligent Logistics Assistant), 90, 185, 186
  IN-ATE, 227
  Intelligent Manufacturing System (IMS), 25, 27
  INTERISK, 115
  KATE, 58, 59
  Knowledge-Lens, 88
  LES, 227
  Maintenance Administration Expert (MAX), 70
  Manufacturing Operations Consultant, 87
  MARK, 79
  Master Scheduling Unit (MSU), 184, 185
  Materials Selection Expert System, 58
  MDX, 227
  Mentor, 93
  MIND, 227
  MUDMAN, 115
  Mycin, 94
  National Dispatch Router (NDR), 190
  Norton Drilling Advisor, 115
  O-PLAN, 224
  PIMS Advisor, 57
  PROSPECTOR, 116
  RAFFLES, 227
  RECONSIDER, 227
  Residence Expert System, 69
  RS/Expert, 72
  Safety Evaluation Expert System (SEES), 59
  SCOPE, 85
  SIGHTPLAN, 224
  SOLO, 224
  TATR, 59
  Vax Performance Advisor, 88
  WELDER, 58
Exxon, 59

Fabrication plant, mechanical, cost in, 3
Factory automation applications, 26
Factory costs, indirect, 3
Factory floor, 43
Factory planning, 181
Failure Diagnosis, Isolation and Recovery (FDIR), 228
Fault isolation, 52, 53
Fiat, 29, 124
Flexibility, 60
Flexible Manufacturing Systems (FMS), 49
Ford, Henry, 124, 182, 240
Ford Motor Company, 142, 168, 191, 226, 258, 279
Forecasting, 177
Forecasting requirements, 78, 79
Fujitsu, 12
Function modeling, 164
Fuzzy engineering, 106, 113, 169, 171, 187, 188
  mathematical principles of, 107
Fuzzy environment, 84
Fuzzy inference programs, 107
Fuzzy logic, 106, 108, 109, 113
Fuzzy relational equations, 108

General Electric, 32, 34, 49, 88–90, 92, 125, 162
General Motors, 18–20, 23, 168, 191, 226, 261, 264
General Telephone and Electronics (GTE), 79, 80, 246
Goldman Sachs, 266
Graphic control frames, 66
GTE Labs, 80

Hardware design, revised, 60
Harvard Business School, 182
Heuristic approaches, 187
Heuristic reasoning, 197
Heuristics, 76
Hewlett-Packard, 23, 245, 281
High-speed Ground Navigation Console (HSGNC), 62
High technology, 29
Hitachi, 12, 57, 64, 66
Homogeneity, 80

Honda, 142
Honeywell, 93
Hughes Aircraft Company, 60, 213
Hypermedia approaches, 114

IBM, 12, 48, 51, 99–104, 182, 227, 244, 280
IBM Japan, 99
Iconic analysis, 175
Idaho National Engineering Laboratory, 84
Idea databases, 118
Image acquisition, 173
Image analysis, 173
Image interpretation, 173
Imperial Chemical Industries (ICI), 53
Inasaka, Fujio, 169
Industrial automation systems, 22
Inference Corporation, 62, 63
Information architecture, 281
Information network, intelligent, 17
Information system, integrated, 15
Information technology, 3, 8, 40
Inquiry
  fuzzy, 107
  quick, 107
Instrument automation, 84
Integrative architecture, 12
Intelligent Business System (IBUS), 90
Intelligent Network (INET), 90
Intelligent networks, any-to-any, 30
Intelligent programming, 20
Intelligent Scheduling Assistant (ISA), 90
Interactive video, 102
Interactivity, 60
Interfaces, natural language, 66
Interfacing, 38
Interference, 38, 67
Internal Revenue Service (IRS), 69
Ishikawajima-Harima Heavy Industries, 132, 226, 227

Johnson Space Center, 62
Just-in-time (JIT) system, 16, 38

Kansai engineering, 234
Kawasaki, 23
Kayaba Industries, 164, 165, 167
Kennedy Space Center, 84, 85
Knowledge
  a priori, 173
  declarative, 117
  procedural, 117
Knowledgebank, 64, 73, 98, 126, 173, 203, 230
  contents of, 66
Knowledgebank management system (KBMS), 119

Knowledge-based construct, 44
Knowledge-based modules, 165
Knowledge-Based Test Assistant (KBTA), 91
Knowledge engineering, 16, 26, 56, 189, 234, 274, 281, 286
  contributions of, 43
  solving aerospace problems through use of, 61
Knowledge engineers, 136
Knowledge management, 44, 45
Knowledge Network, 88
Knowledge processing, 45
Knowledge processing systems, practical use of, 54
Knowledge representation scheme, in expert system, 157
Knowledge-representation structures, 127
Knowledge transmission, 270
Kodak, 154
Komatsu Corp. 131
Kono, S. 134
Kurosu, Kenji, 169

Laboratory of International Fuzzy Engineering, 169
Laboratory research, AI technology in, 52
Lagrangian relaxation, 202
Language constructs, 100
Large scale integration, 155
Lawrence Livermore National Laboratory (LLNL), 81–84
Linear Programming (LP), 186
Loading, 180
Logical consistency, 166

Machine learning mechanisms, 114
Macroengineering, 72
Macroengineering concepts, 72
Magnetic fusion energy (MFE), 81
Manager, manufacturing industry, 6
Man's Physical and Mental Work, milestone developments in, 8
Manufacturing, applications in, 50
Manufacturing Automation protocol (MAP), 264
Manufacturing discipline, new, 6
Manufacturing engineering, 6
Manufacturing management, 5, 201, 202
Manufacturing management system, integrative, 260
Manufacturing Process Planning System (MPPS), 191
Marketing, 237, 239
Martin Marietta Aerospace, 53, 131
Massachusetts Institute of Technology, 128, 142
Masui, S. 134
Mayo, Eton, 182
McDermott, Prof. 94
McNamara principle, 78

Messerschmidt, Boelkow, Blohm (MBB), 227–30
Metadata, 119
Metaknowledge, 117, 158
Metaknowledge engineering discipline, 276
Metalevel processing, 192
Metapatterns, 113
Method, 112
Micromachines, 128
Microprocessor sensor, 128
Million instructions per second (MIPS), 7
Ministry of International Trade and Industry (MITI), 27
Mitsubishi Heavy Industries, 227
Model, analytical, 73, 166
Modeling, 51
　function, 164
　logic, 164
　physical, 164
Models, 76
Models, mathematical, 73, 77, 147
　computer-based, 196
Monitored Decision Script, 83
Monsanto Chemicals, 258
Mrs. Fields Cookies, 244
Multidimensional system, 23
Multimedia approaches, managing, 64
Multimedia information storage, 164
Multivendor Integration Architecture (MIA), 12
Murayama, Yojiro, 169

NASA, 50, 59, 62, 106, 131, 191
National Institute of Standards and Technology (formerly NBS), 68
Natural language facilities, 66
Navistar International, knowledge engineering at, 63, 64
NEC, 12
Network analysis, electronic, 70
Network design, 72, 75
　process, 74
Networking, 144
Network management, 79
Networks, distributed, designers of, 78
Network Troubleshooting Consulting, 90
Neural network circuit, 110
Neural network projects, in manufacturing, 110
Neural networks, 113
Neural networks applications, in mathematics, 110
Neural system, knowledge of, 111
Nexpert Object, 57
Nissan Motors, 23, 125, 226, 227
Northrop Corp. 162
NTT, 12, 15

Numerical control (NC), 267
NYNEX, 70

Object-oriented design methods, 12
Object processing, 204
Object recognition, 127
Odetics, 132
Olivetti, 262
Onboard computers, 168
Onboard electronics, 168
Online consultation, 102
Online diagnostics, AI-enriched, 168
Operating characteristics (OC) curves, 112, 215
Optical disks, 29
Optimization, 149, 223
Optimizers, 18
Output configuration, evaluation of, 97

Packet-switching, 79
Packet-switching communications, topological design for, 74
Packet-switching nodes (PSNs), 75
Parameterization, 223
Parametric approaches, 221
Pattern analysis, 114, 117
　computer-based, 116
Pattern matching, 115
Personal computer, labor content of, 3
Philips, 237, 246
Plant Performance Monitoring Systems (PPMS), 85
Possibility theory, 188, 197
Preliminary construction sequencer, 57
Principal Component Analysis (PCA), 134
Procedure Qualification Record (PQR), 58
Process design, 27
Processes, sequential, 41
Product designers, 223
Production
　automated, 213
　lean, 24
Production control system, 178, 210
Production management, 177, 199
Production plan, 199
Production planning, 43, 177
Production planning model, 188
Profitability analysis, 253
Profit centers, 263, 264
Programming language, object-oriented, 73
Project Appraisal Resourcing and Control (PARC), 68
Project R1, 94
　R1/XCON, 95
Project status, 259, 260
PROPHET, 79

Protocol model, application of, 74
Protocols, 72
Prototype, 67, 68, 96
Prototyping, 118, 223

Quality circles, 182
Quality control (QC), 63, 135, 213
    processes, 4
Query analysis and meaning-extraction frames, 66
Query optimizer, 119
Queueing theory, 41
Quotron, 240

RAND Corporation, 59
Reactor Emergency Action Level Monitor (REALM), 85, 86
Realtime hybrid system solutions, 72
Reasoning, evidential, 227, 228
Reasoning methods, 161
Recognition knowledge, 92
Regression testing, 97
Rensselaer Polytechnic Institute, 59
Research prototype, 99
Resource allocation, 209
Resource groups, 18
Ribout, Jean, 116
Robotics, 4, 125, 280
    advanced, 126
Robots, 122, 125–29
    field, 129, 130
    intelligent, 123
    moving, 130
Rocket scientists, 32
Rockwell International, 27
Routing, 180, 211
Royal Dutch/Shell, 16
Rule-based systems, 120
Rule chip, 106
Rule language, 247
Rules, 93

SAAB, 53
Saturn project, 18, 19, 23, 24, 65
Scalar space, 114
Scheduling, 87, 180, 196, 213
Scheduling expert system, 211
Scheduling systems, automatic, 4
Schlumberger, 116, 153, 154
Search heuristics, 65
Sears, Richard W. 241
Sears Roebuck, 182
Security administrator, 283
Security architecture, 283

Security risk, 284
Segmentation, 173
Semantic modeling, 119
Shaw, George Bernard, 14
Shewhart QC chart, 215
Shewhart, Walter, 215
Siemens, 262
Simulation, 43, 148, 211
Simulation and artificial intelligence systems, 78
Simulation languages, 60
Simulation program, 57
Simulation studies, 40
Simulators, 16, 18, 24, 248
Simultaneous engineering, 167
Sixth generation computers (6GC), 136, 137
Skillet line, 24
Software, 96
    advanced, 38
Software inventory and reuse expert systems, 191
Software solutions, new, 12
Sohio, 154
Solution selling, 106
Sorell, Louis Prof. 235
Source conditioning, 82
Sperry, 236
Stanford University, 94
Startup utility, smart, 103
Strategic plan, 4, 5
Strategic planning, 4, 5
State University of New York, 89
Structure of command, 33
Subject map frames, 66
Sumitomo Electric, 146
Suntory, 146
Supercomputers, 44, 120
Sustenance, 97, 100, 104
Switch placement problems, 77
Symbolic computing, 159
Symbolic processing, 175
System architect, 31
System architecture, 31, 32
System management, 155
Systems, cooperative, 18
Systems responsibility, 26

TAI, 84–86
Tamaki, Hiroya, 169
Task management, 75
Technical and Office Protocol (TOP), 264
Teijin Systems Technology, 146
Terano, Toshiro Prof. 108, 134, 169, 172
Testing, 51, 149
Thinking Machines, 7

Tokyo Institute of Technology, 108
Tokyo University, 128
Toppan Printing, 146
Toshiba, 131, 132
Total quality management (TQM), 49
Touch sensing, 135
Toyota City, 128
Toyota Machine Works, 146
Toyota Motor Corp. 29, 128, 168, 226, 241, 258
Tramiel, Jack, 233, 235, 237
Turnaround time, 41, 280

Uncertainty, 196
Unisys, 236
United Airlines, 48
United Auto Workers Union, 18
United Parcel Service (UPS), 244
U.S. Army, 50
U.S. Navy, 60
United States Nuclear Regulatory Commission (USNRC), 81
United Technologies, 27
University of California, at Berkeley, 128
User interface, 60

Value differentiation, 106
Vamos, Tibor Dr. 114
Vector space, 114
Vision systems, 135
Visualization, 42
VMS Expert (TVX), 91
Volkswagen, 168
Volvo, 23

Warehouse placement problems, 77
Waterfall model, 12
Watson Sr. Thomas, 182
Westinghouse Electric, 132, 184, 218
Wood, Robert E. 182
Workspace definition, 126
Worldwide Engineering Release System (WERS), 279, 288

Xerox, 16, 48

Yamamoto, 134
Yorktown Expert System for MVS (YES/MVS), 99, 100
Yoshikawa, Hiroyuki Prof. 25